Copyright © 2024 Antonio M.

All rights reserved. No part of this publication may be reproduced or transmitted in any form or by any means, electronic or mechanical, including photocopying and recording, without the prior written consent of the author or publisher.

Title: Mechanics in Motion: Exploring the Technological Heart of Modern Vehicles

Author: Antonio Marino

Legal notice: Copyright violations will be prosecuted.

Mechanics in Motion: Exploring the Technological Heart of Modern Vehicles

Index

Mechanics in Motion: Exploring the Technological Heart of Modern Vehicles — 2

Introduction: A Journey Through the Mechanics and Sustainability of Modern Vehicles — 6

Chapter 1: Introduction to Modern Vehicle Mechanics — 11

History, Fundamental Concepts and Propulsion of the Future — 11

Fundamental Concepts of Automotive Mechanics — 14

Introduction to modern propulsion systems — 17

Chapter 2: Vehicle Structure and Components — 21

Description of the structure of a vehicle — 22

Analysis of the Main Components — 25

Chapter 3: Vehicle Maintenance and Care — 29

Practical Tips for Regular Vehicle Maintenance — 30

Procedures for Checking and Replacing Wear Parts — 33

Importance of Technical Assistance and Periodic Overhauls — 40

Useful Accessories to Always Keep with You in the Car — 42

Chapter 4: Types of Nutrition — 45

Insight into the Various Types of Engines — 45

Difference Analysis Between Power Systems — 48
- Internal Combustion Engines - Petrol and Diesel — 49
- The Battle of Ibridi — 50
- Electric Vehicles — 51
- Evolutionary Trends and Technical Impact — 51

Advantages and Disadvantages of Automotive Powertrain Technologies — 53
- Internal Combustion Engines, Petrol and Diesel — 53
- The Battle of Ibridi — 55
- Electric Vehicles — 57

Chapter 5: Motorsport Technologies, Analysis of Different Existing Solutions — 59

Internal Combustion Engines: Petrol and Diesel — 60
- Differences between Petrol and Diesel Engines — 60
- Technological Evolution — 63

Hybrid Engines: A Journey Through Parallel Configurations, Series and Other Technologies — 70
- Parallel Hybrids — 70
- Hybrids Series — 72
- Plug-in hybrids (PHEV) — 74
- Mild Hybrid — 76
- Diesel hybrids — 78
- Compressed Air Hybrids — 79

Hydrogen Engines: Hydrogen Engines Exploration — 81
- Fuel Cell Operation — 84

Advanced Electric Motors: Batteries, Motors and Emerging Technologies — 86
- Permanent Magnet Motors: High Efficiency and Performance — 86
- Induction Motors: Simplicity and Reliability in Tesla Model S — 89
- Variable Reluctance Synchronous Motors: Efficiency in Jaguar I-PACE Vehicles — 92
- Linear Motors: Advanced Propulsion in High-Speed Maglev Trains — 95
- Axial Flow Motors: Optimized Performance in Lucid Air Electric Vehicles — 98

High Capacity and Power Batteries — 100
- Lithium-Ion (Li-Ion) Batteries — 101
- Lithium Iron Phosphate (LiFePO4) Batteries — 102
- Lithium Titanate (LTO) batteries — 103
- Batteries of the Future — 104

Chapter 6: Transmission Systems — 106

Manual Transmission — 106

Automatic Transmission — 108

Dual Clutch Transmission (DCT) — 111

Continuously variable transmission (CVT) — 113

All-Wheel Drive (AWD) and Four Wheel Drive (4WD) — 115

Semi-automatic transmission — 117

Hybrid Transmission 119

Electric Transmission 120

Chapter 7: Automotive Safety 123

Advanced Driver Assistance Technologies (ADAS) 123

Automatic Emergency Braking Systems (AEB) 126

Smart Roads and V2X Communication 130

Impact of Autonomous Driving on Safety 134

Advanced Driver Monitoring Systems 137

Chapter 8: A Planet in Balance, Eco-Sustainable Vehicles and the Fight against Pollution 140

Environmental Impact of Internal Combustion Vehicles 141

Solutions to Reduce the Environmental Impact of Internal Combustion Vehicles 142
- Development of Sustainable Fuels 143
- Emission Reduction Solutions 146
- Vehicle Weight Reduction 150

Detailed Comparison in Emissions, Resource Consumption and Sustainability 153

Conclusion 157

Introduction: A Journey Through the Mechanics and Sustainability of Modern Vehicles

Welcome to an exciting journey through the world of modern automotive mechanics and its incessant evolution. This book aims to open the doors of knowledge, inviting you to explore the beating heart of contemporary vehicles.

From a historical overview and fundamental concepts to the technologies of the future, this guide offers a comprehensive perspective on the complexity and beauty behind the wheel of every vehicle. Without going into the details of the chapters, we will immerse ourselves in a world that ranges from the structure to the components of the vehicles, their maintenance, the different types of power supply, transmission and safety systems up to the most advanced engine technologies.

This book is more than just a guide; It is a travel companion that will take you through the roads of automotive mechanics, giving you a clear and accessible perspective on complex topics. Whether you're passionate about engines, technology or sustainability, get ready for an enlightening and engaging journey into the world of modern vehicles.

Chapter 1: Introduction to Modern Vehicle Mechanics

We will begin our journey with a historical overview, immersing ourselves in the first mechanical motions that laid the foundations of the modern automotive world. From those early pioneering dreams, we'll move through the decades to understand how automotive mechanics have changed over time.

Chapter 2: Vehicle Structure and Components

In the second chapter, we will explore the complex structure of a modern vehicle. We'll reveal the secrets of the engine roaring under the bonnet, analyse the driving force behind the drivetrain and touch on every element that synergistically makes our daily journey possible.

Chapter 3: Vehicle Maintenance and Care

The third chapter will guide us through crucial maintenance practices, ensuring that our faithful companion on four wheels always remains at peak performance. From engine care to attention to essential components, we will find out how to ensure the longevity and efficiency of our vehicle.

Chapter 4: Types of Nutrition

In chapter four, we will address the revolution in power systems. From classic internal combustion engines to hybrid and electric innovation, we will explore powertrain options and unveil the different paths to sustainability.

Chapter 5: Motorsport Technologies, Analysis of Different Existing Solutions

Within this chapter, we will dive into an in-depth exploration of motorsport technologies, outlining the different souls that drive innovation in the automotive sector. From consolidating internal combustion engines to sophisticated hybrid propulsion and progressive electric vehicles.

Chapter 6: Transmission Systems

In this chapter, we explore the complex powertrain technologies that drive modern vehicles. From classic manual transmissions to automatic transmissions to hybrid and electric solutions, we look at the advantages, disadvantages and the future of automotive engineering.

Chapter 7: Automotive Safety

The penultimate chapter explores advanced automotive safety technologies, such as automatic braking systems, driver assistance systems, sensors in general, and more.

Chapter 8: A Planet in Balance, Eco-Sustainable Vehicles and the Fight against Pollution

Finally, we'll end our journey with an in-depth look at the environmental impact of vehicles. We will explore eco-sustainable alternatives, from the growth of hybrids to the electric revolution, and we will openly address the challenge of vehicular pollution, looking for solutions for a cleaner and more sustainable future.

Get ready to dive into the depths of modern automotive mechanics, where power blends with efficiency and innovation paves the way for an eco-friendly journey. Be ready to guide through the pages of this book, where science meets passion and each chapter unveils new secrets under the hood. Welcome to a world of knowledge, power, and boundless guidance.

Chapter 1: Introduction to Modern Vehicle Mechanics

Chapter 1 takes us on a journey through the history, fundamental concepts and basics of vehicle propulsion. We will explore the evolution of means of transport, from its origins to the present day, focusing on the key principles that guide the operation of modern vehicles. With a generic approach, we will try to make the basic concepts accessible, opening the door to a fascinating world of innovation and technology.

History, Fundamental Concepts and Propulsion of the Future

Along the avenue of automotive history, the wheels of technology have traced a once-in-a-lifetime path, intertwining the past with the present and opening the doors to the future. The history of vehicles is a fascinating tale of inventions, challenges overcome and daring innovations that have transformed the concept of mobility.

The roots of this narrative lie in the dreams of pioneering visionaries, who dared to imagine a world without the reins of the horse. From the first experiments with steam vehicles in the eighteenth (eighteenth century), to Karl Benz's revolutionary petrol automobile in 1885, automotive history is a succession of adventurous steps. Every tiring curve of the route has led to new horizons, opening roads never travelled and rewriting the destiny of mobility.

But what makes the miracle of automotive mechanics possible? That's where the fundamental concepts come in. In the jumble of gears, pistons and belts, we will address the pillars of vehicular mechanics. The transmission, the beating heart of driving dynamics, merges with the suspension, guaranteeing a smooth ride on uneven roads. Delving into the operation of the engine, we will discover the synchronized ballet of combustion and energy, a symphony of power under the hood.

Over the years, the complex mosaic of automotive mechanics has embraced new challenges and evolved. The future of propulsion is already among us, an extraordinary chapter in a book that is still being written. In this race towards sustainability, modern automotive propulsion is taking on new forms.

Modern propulsion systems, ranging from hybrid engines that harness the synergy between tradition and electricity, to all-electric vehicles embracing the green revolution, shape the future of our journey. The road to an era of sustainable mobility is being paved by hydrogen-powered vehicles, plug-in hybrid vehicles, and new technologies emerging from the minds of the boldest engineers.

In conclusion, history, fundamental concepts and the propulsion of the future are intertwined in a never-ending journey, where evolution and innovation are the turning point. From a pioneering phase of bold dreams, we are projecting ourselves into the future, driven by the same thirst for innovation that took the first steps of Karl Benz and Henry Ford. The history of vehicles is a fascinating odyssey, and the road ahead of us is still long and open to new discoveries.

Fundamental Concepts of Automotive Mechanics

At the beating heart of every vehicle, among the twists and turns of intricate components, lie the fundamental concepts that form the backbone of modern automotive mechanics. It's like opening an ancient engineering book, where each page would reveal an essential chapter in the story of how a vehicle comes to life and moves through space.

A fundamental element that has driven the automotive revolution is undoubtedly the transmission, the dynamic heart that translates the power of the engine into motion. From gears to modern automatic transmissions, this component is the master of synchrony, adjusting gears to suit different road and driving conditions. The broadcast is the soundtrack that accompanies the ballet of the wheels on paved roads.

Alongside the transmission, the suspension acts as the "comforter" of the ride, absorbing shocks and ensuring a smooth ride. Springs, shock absorbers and anti-roll bars work together in perfect synergy to provide stability and comfort for the driver and passengers.

Suspension is the key to maintaining wheel contact with the road, which is crucial for safe and efficient driving.

Moving on to the very soul of the vehicle, we immerse ourselves in the engine, the vital organ of every car. Internal combustion is the choreographed dance of pistons, cylinders and valves that converts fuel energy into power. From traditional internal combustion engines to the latest turbo and hybrid engines, the constant quest for efficiency and performance drives engineers to experiment and innovate.

Vehicle dynamics is a science, a fusion of physics and engineering that governs the behaviour of a moving vehicle. The center of gravity, lateral acceleration and weight distribution are just some of the variables that engineers must balance to ensure a stable and safe ride. A well-designed vehicle is like an agile athlete that moves with precision and control.

Pushing the vehicle forward, like a powerful voice in the orchestra, we find the fuel system. From petrol or diesel-powered engines to new alternative technologies, the way a vehicle feeds on energy are a direct reflection of environmental challenges and the growing demand for sustainability.

The transition to electric vehicles is revolutionizing the automotive landscape, bringing with it promises of lower emissions and greater energy efficiency.

Finally, brake technology concludes the mechanical ballet. From primitive drum braking systems to modern brake discs and anti-lock braking systems (ABS), speed control is just as important as the technological advancement that has led to such innovations. Brakes are the guardian of safety, and their evolution reflects the constant commitment to improve effectiveness and responsiveness in emergency situations.

In conclusion, the fundamental concepts of automotive mechanics are the pillars on which the entire edifice of mobility is built. In the constant flow of innovation, the future promises further fascinating chapters, in which these fundamental concepts will be redefined and expanded, pushing the entire industry towards ever more ambitious horizons.

Introduction to modern propulsion systems

The mobility horizon is constantly evolving, shaped by human ingenuity and growing environmental awareness. In this era of transformation, the introduction to modern propulsion systems emerges as an intricate web of innovation, sustainability and technological progress.

This chapter aims to explore the complex landscape of automotive propulsion, from the ancient mastery of the internal combustion engine to the fascinating world of emerging technologies. The internal combustion vehicle, a symbol of a bygone era, has long been the undisputed star of the roads. However, its supremacy has been challenged by growing environmental awareness and the need to reduce harmful emissions.

Electrification has taken center stage, with electric vehicles (EVs) emerging as the protagonists of the new era. Electric vehicles are powered by advanced batteries that provide the energy needed to power an electric motor.

This technology promises to revolutionize mobility, offering a clean and sustainable alternative to the traditional internal combustion engine. The exponential growth in battery capacity and decreasing costs are making EVs increasingly accessible, paving the way for a massive transition to electric mobility.

In an effort to mediate between tradition and innovation, hybrid vehicles have gained popularity. These vehicles combine an internal combustion engine with an electric motor, taking advantage of the best of both worlds. The ability to run on both conventional fuel and electric power provides flexibility that adapts to different driver needs.

Hybrids are an intermediate step, a technological bridge that facilitates the transition to all-electric mobility. In addition to electrification, the search for alternative fuels is gaining traction. Hydrogen presents itself as a promising clean energy resource. In fuel cell vehicles, hydrogen reacts with oxygen in the air to generate electricity, powering the electric motor and producing only water as a byproduct. This technology paves the way for a future where the vehicle is not only sustainable during use, but also in the production and distribution of the hydrogen itself. Despite the exciting promises, the large-scale adoption of electric vehicles and other advanced propulsion technologies comes with significant challenges. The issue of charging infrastructure remains one of the main obstacles. Building a widespread charging network is key to ensuring the convenience and efficiency of electric vehicles, especially in an environment where demand is set to grow.

Another challenge is the durability and sustainability of the batteries. While battery technologies are making remarkable progress, there is a need to address the issue of recycling and responsible disposal of spent batteries. The industry needs to invest in research and development to improve the lifespan of batteries and make their life cycle more sustainable. Finally, the issue of costs remains a deterrent for many consumers.

Although the prices of EV technologies are gradually decreasing, the initial cost gap compared to traditional vehicles is still a barrier for some buyers. Government incentives and continuous innovation are key to reducing this disparity and accelerating the adoption of more sustainable vehicles. As we dive into the waters of electrification, further scenarios emerge for the future of automotive mobility.

Autonomous driving is presented as a key element, with vehicles capable of traveling intelligently and safely through the roads without direct human intervention. The integration of autonomous driving systems is a significant step towards increased road safety and increased traffic efficiency. Advanced connectivity is another crucial element. Vehicles that communicate with each other and with the surrounding infrastructure promise to optimize traffic flow, reduce accidents and improve the overall driving experience.

The synergy between vehicles and smart infrastructure paves the way for smoother and more connected mobility. Continued research into the innovation of lightweight materials and the technologies mentioned above is helping to make vehicles more fuel-efficient, reducing the overall environmental footprint. Advanced materials such as carbon fiber and aluminum are gradually replacing traditional steel components, reducing the weight of vehicles and improving energy efficiency.

The introduction to modern drive systems is a fascinating perspective on the future of mobility. Electrification, along with brilliant new technologies and solutions, promises to redefine the way we get around. Despite the challenges, the automotive industry is demonstrating a significant commitment to sustainability and innovation. The road to cleaner, smarter mobility is open, and the journey continues to drive the future of automotive powertrain.

Chapter 2: Vehicle Structure and Components

Before delving into the complexity of the structure of a modern vehicle, it is essential to draw an introductory framework that leads us through the intricate labyrinth of its components.

The detailed description of a vehicle's structure is like a map that guides us through the intricate meanders of its design, revealing the secrets of each element that contributes to the harmonious functioning of this machine on four wheels.

Imagine yourself as an explorer, ready to uncover the mysteries hidden under the hood and behind each panel, as we dive into the extraordinary engineering that brings the modern vehicle to life. With this adventurous spirit, we now enter our exploration with the "Description of the structure of a vehicle".

Description of the structure of a vehicle

Describing a vehicle reveals a rich and complex landscape, but to fully understand the modern automotive ecosystem, it is crucial to examine the pros and cons of the different technologies employed. From the chassis to the engine, from the transmission to the electronics, every component has implications that go beyond its apparent function, contributing to both the positive and negative aspects of the driving experience and environmental impact.

Let's start by analyzing the chassis, the load-bearing structure of the vehicle. Traditional ladder and crossmember frames offer a robust and durable structure, ideal for heavy-duty vehicles such as trucks and off-road vehicles. However, these frames can be heavier and less fuel-efficient than modern monocoque frames, which are lighter and offer greater structural rigidity.

The lightweight advantage results in greater fuel efficiency and more agile performance, but this may result in lower impact resistance in the event of a collision compared to more traditional frames.

Moving on to the engine, the choice between internal combustion engines and electric motors involves a number of considerations. Internal combustion engines are known for their reliability and extended range, but they emit harmful gases and are less energy efficient than electric motors. Electric vehicles, powered by batteries, are greener and offer more immediate acceleration. However, their range can be limited, and the production and disposal of batteries raise various environmental concerns. In addition, energy production from low-emission sources is crucial to maximising the positive impact of electric vehicles.

The transmission directly affects the performance and efficiency of the vehicle. Manual transmissions offer more control from the driver, but they can require a learning curve and can be less efficient than automatic transmissions, especially in urban traffic conditions. CVT (continuously variable) transmissions seek to optimise the gear ratio continuously, offering a compromise between efficiency and smooth driving.

The electronics integrated into modern vehicles are a technological double blade, improving safety and introducing new conveniences, but also raising concerns in terms of reliability and cyber vulnerability. Driver assistance systems, such as stability control and automatic emergency braking, improve road safety. On the other hand, the increasing reliance on electronics paves the way for cybersecurity threats and potential complex failures that require specialized skills to repair.

In the context of suspension and wheels, different configurations offer specific advantages. Air suspension provides a more comfortable ride and the ability to adjust the height of the vehicle, but it can be expensive to repair in the event of a breakdown. Light alloy wheels improve aesthetic appeal and reduce unsprung weight, improving handling, but they can be more expensive to replace than steel wheels.

Finally, the vehicle's body and aesthetics play an essential role in attracting consumers, but also in influencing aerodynamics and fuel efficiency. SUVs can provide a robust design and high ride height, but they often sacrifice fuel efficiency compared to more aerodynamic sedans.

The focus on exterior design must balance aesthetics with functionality, ensuring that the vehicle's attractive appearance does not compromise its efficiency.

In conclusion, choosing between different technologies and components in a vehicle's structure involves a number of trade-offs. Consumers must weigh the benefits in terms of performance, fuel efficiency, safety, and environmental impact against the potential drawbacks. As the automotive industry continues to evolve towards increasingly sustainable and technologically advanced solutions, a thorough understanding of these pros and cons is crucial to driving informed choices for both consumers and vehicle designers.

Analysis of the Main Components

In the context of the analysis of the main automotive components, we explore in detail the operation, characteristics and differences between the various technologies, focusing on brands, models and comparisons that can affect performance, consumption and other relevant aspects.

Engine:

The engine is the beating heart of a vehicle, and the choice between internal combustion engines and electric motors greatly defines the driving experience. For example, iconic vehicles such as the Toyota Prius helped popularize hybrid vehicles, combining an internal combustion engine with an electric motor. All-electric cars, such as the Tesla Model S, have gained attention for their immediate acceleration and zero emissions while driving.

However, limited range and concerns about charging can impact consumer choices. Internal combustion vehicles maintain a loyal following thanks to their comfortable and sporty performance, as well as the presence of an established network of filling stations. Innovations such as hydrogen engines, pioneered by brands such as the Toyota Mirai, offer a promising alternative with zero emissions during use.

Transmission:

The drivetrain is crucial for the efficient propagation of power to the wheels. Sports models, such as the Porsche 911 with its PDK (Porsche Dual Clutch) transmission, offer ultra-fast gear changes, improving both performance and efficiency. Cars equipped with CVT transmissions, such as the Honda Accord, try to keep the engine in its optimal power range, improving fuel consumption. In contrast, manual transmissions, found in the Mazda MX-5, for example, can offer greater driver involvement but require more attention in urban traffic. Conventional automatic transmissions, such as those in the Mercedes-Benz S-Class, provide superior comfort during urban driving and over long distances, at the expense of sporty involvement.

Loom:

The chassis plays a crucial role in the stability and safety of the vehicle. Traditional chassis, such as those used by the Ford F-150, offer strength and durability, suitable for heavy-duty vehicles and towing. On the other hand, the monocoque frames, found in the Honda Civic, reduce the overall weight of the vehicle, improving fuel efficiency and handling. Luxury brands, such as BMW with its "Carbon Core" approach, integrate lightweight and durable materials to improve performance and safety. Sports cars, such as the Chevrolet Corvette, often use aluminum frames to reduce weight and improve driving dynamics.

Suspension systems:

Suspension systems influence driving comfort, stability and ease of handling the vehicle. Air suspension, such as that offered by Mercedes-Benz on the S-Class, allows for continuous adjustment of the vehicle height and a more comfortable ride. Conversely, sports suspension, such as that found on the Porsche Cayman, can offer greater steering precision and cornering grip, at the expense of comfort on rough roads.

Off-road vehicles, such as the Jeep Wrangler, often adopt stronger suspension to tackle rough terrain. Magnetorheological (magnetic fluid) suspension technology, such as that implemented on the Chevrolet Camaro, allows instant adjustments to suit driving conditions.

Brakes:

Braking systems are crucial for vehicle safety and performance. Disc brakes, like those found on many high-end cars, offer increased braking power and heat dissipation. However, drum brakes, still used on some compact vehicles, can be cheaper and require less maintenance. Regenerative brakes, common in electric vehicles such as the Nissan Leaf, recover energy during decelerations, improving efficiency and reducing wear and tear on traditional components.

In summary, the choice of key components within a vehicle is closely linked to the driver's needs, preferred type of driving and environmental considerations. Makes and models offer a wide range of options, each with its own pros and cons, creating a diverse and ever-changing automotive market.

Chapter 3: Vehicle Maintenance and Care

In the vast theater of automotive mechanics, the stage of vehicle maintenance and care stands as a crucial act, where gears and various components are intertwined with the dedication of the owner. Welcome to a chapter where a passion for riding merges with a commitment to maintenance, an ode to the longevity and efficiency of our faithful companion on four wheels.

As custodians of these mechanical marvels, we are called upon to dance with clarity among the chrome details of maintenance, preserving the vitality of our vehicle. On this journey, we'll explore the intricate rituals that keep our vehicle in harmony with the roads, from the delicate touch of control routines to the deeper art of repair and replacement. Let's get ready to immerse ourselves in the choreography of automotive care, where our attention becomes the key to a smooth journey without unexpected surprises.

Practical Tips for Regular Vehicle Maintenance

In the vast context of automotive mechanics, regular maintenance is configured as a fundamental pillar to preserve the health and efficiency of our means of transport. Every component, no matter how small it may be, plays a crucial role in the overall car, and underestimating its care can result in a decline in performance and safety on the road.

Neglected components act as elements that can disturb the harmony of the machine or cause problems with its proper functioning, potentially compromising the driving experience and jeopardizing the safety of the driver and passengers. Worn brakes, for example, can extend braking distances, increasing the risk of accidents. A neglected suspension system can result in an unsteady ride, affecting the vehicle's stability over rough terrain. Without forgetting the crucial importance of oils and fluids, which are essential to ensure the proper functioning and durability of mechanical elements.

In this intricate balance of engineering, the engine plays a central role, and each type of propulsion requires specific care. In internal combustion vehicles, attentions such as regular cleaning of the injectors and periodic replacement of spark plugs are real refrains of care, essential to maintain engine efficiency and reduce harmful emissions. For electric vehicles, careful battery management is the focus, as constantly monitoring battery capacity and health is crucial to ensure a smooth driving experience.

It's important to note that while EVs generally require less maintenance as far as the engine is concerned, replacing a battery can incur significant costs. However, as technology continues to evolve and market competitiveness grows, the costs associated with batteries are gradually decreasing, making EVs more accessible and affordable in the long run.

Regular maintenance is a direct investment in the safety, efficiency and longevity of your vehicle. A preventive strategy, based on periodic inspections and timely interventions, not only reduces the risk of sudden breakdowns, but also preserves the residual value of the vehicle over time. Exploring the maintenance cost landscape requires careful consideration, considering variables such as vehicle model, engine and engine and transmission make and type.

In the comparison between electric vehicles, internal combustion vehicles and hybrid technologies, differences emerge both in terms of frequency of interventions and in cost management. While electric vehicles may require less maintenance for the engine, battery replacement is something that needs to be carefully considered. Internal combustion vehicles, on the other hand, may involve more frequent expenses, but spread over a wider range of components. Hybrid technologies, intermediate between the two, present a maintenance framework that combines elements of the two technologies, offering a balance between efficiency and operating costs.

Procedures for Checking and Replacing Wear Parts

Within the complex mechanics of our vehicle, certain parts play a crucial role in ensuring optimal performance, but they are also exposed to a natural wear process. To preserve the efficiency and safety of the vehicle, it is essential to adopt periodic control procedures and, when necessary, proceed with the replacement of these key components.

Discs and Brakes:

The braking system, as we have already said, is essential for the safety of the vehicle. Regular inspection procedures include checking the thickness of the pads and discs. Worn pads and excessively worn discs can impair the vehicle's ability to stop, making timely replacement crucial to ensure safe operation. In addition, regular checking of the brake fluid level and quality is essential to ensure proper functioning of the system. The hydroscopic nature of brake fluid causes it to gradually absorb water into the brake system. This process can compromise the effectiveness of the brake fluid, reducing the performance of the brake system and increasing the risk of corrosion. However, it is essential to monitor the boiling point of the fluid and replace it regularly to maintain the optimal performance of the brake system (in general, many car manufacturers suggest a replacement every 1-2 years or every 40,000-50,000 km, whichever comes first).

Tyres:

Tyres, as a point of direct contact with the ground, are exposed to uninterrupted wear. Regular control of tread depth is essential to ensure optimal grip on different road surfaces. Worn tires increase the risk of slips and accidents. In addition, assessing tire pressure contributes not only to safety, but also to fuel efficiency. It is important to note that incorrect wheel alignment can affect tire wear, leading to uneven wear. This problem can result from factors such as impacts with potholes, incorrect trim adjustments, or lack of maintenance. Therefore, regular checks on tire trim and wear are crucial to preserve safety and optimize vehicle performance.

Air, Oil and Cabin Filter:

Clean airflow is vital for the proper functioning of the engine. Air filters, over time, accumulate dust and debris that can affect performance. Similarly, the oil filter plays a critical role in engine lubrication. Periodic checks and regular replacements of both filters are practical to ensure the vitality of the engine and extend its life. Engine air filters in vehicles are available in different variants designed to meet specific needs, there are conventional filters, made of materials such as paper or fiberglass, which remove dust and particles from the air entering the engine, high-performance filters, characterized by advanced materials such as cotton or fabrics to optimize airflow and offer greater filtration capacity than traditional filters. In addition, sports filters are designed to improve airflow and engine efficiency, some of which are also suitable for low temperatures, ensuring high performance in colder weather conditions. As for the cabin filter, its regular replacement helps to keep the air inside the cabin free of particles and allergens. In the case of cabin air filters, there are standard, antibacterial/anti-allergic and activated carbon filters, each with features aimed at improving interior air quality. Your filter choice should consider the manufacturer's recommendations and the specifications of your vehicle.

Timing Belts and Accessories:

Belts play a key role in ensuring that the different components of the motor operate in sync. However, over time, they can deteriorate. Checking the tension and integrity of timing and accessory belts are critical checks. Belt breakage can cause serious damage to the engine, making preventive replacement essential (many manufacturers recommend replacing the timing belt every 80,000 km or between 5 and 7 years).

The accessory belt in a vehicle is connected to different components such as the alternator or air conditioning compressor. Its breakage can cause significant problems, including loss of battery power.

In addition, it should be noted that in addition to the timing belt there is the timing chain, which have the same task in terms of functionality, some accessory components such as, for example, the water pump that is connected to the latter or to the service belt this depends on the technical construction of the vehicle.

Battery:

The battery provides the energy needed to start the vehicle and power the electrical systems. An old or damaged battery can cause starting difficulties. Regular battery charge and health checks, as well as replacement, when necessary, keep your vehicle ready for use. The average duration is 4-8 years. The standard voltage is 12 volts. Among the types of batteries, traditional lead-acid batteries require periodic maintenance and are commonly used. AGM (Absorbent Glass Mat) batteries are sealed, offer vibration resistance and require little maintenance. While lithium batteries, which are lighter and longer-lasting, are gaining popularity for their advanced performance. The main differences between AGM, lithium and lead-acid relate to technology, durability, weight and vibration resistance. The choice depends on the needs of the vehicle and the desired performance.

Fuel Filter:

The fuel filter, crucial for the maintenance of diesel, gasoline and gas-powered vehicles, requires regular checks and recommended replacements every 20,000 to 30,000 km, varying according to the fuel type and manufacturer specifications. In the context of gas engines, the frequency of replacement can vary depending on the type of gas used. Accurate handling of this component is critical to optimizing vehicle performance, preventing engine malfunctions, and extending the life of the fuel system. Investing in proper fuel filter maintenance is crucial for safe and efficient driving.

Suspension Components:

Shock absorbers are crucial parts of a vehicle's suspension system. To ensure proper operation, regular inspections are essential. Look for any signs of oil leakage, damage, or corrosion. Replacement of shock absorbers is recommended every 80,000 to 100,000 km or if there are obvious signs of wear. It is important to note that the replacement should take place in pairs, both for the front and rear axles.

Suspension springs, which are responsible for shock absorption, require careful control to detect deformation, breakage or signs of sagging. If they have obvious defects or a loss of height occurs, replacement is recommended. There is no standard interval for substitution, but it is essential to perform it in pairs to maintain a balanced suspension.

Suspension bushings deserve regular inspections for damage or signs of wear. Replacement of bushings is necessary when they show play or signs of deterioration. The replacement interval may vary depending on many parameters.

Anti-roll bars, which are responsible for stabilizing the vehicle during corners, require regular checks for damage or weak connections. If there are obvious defects, it is advisable to replace the bars. Again, there is no standard interval for replacement, but it is important to perform it in pairs for suspension stability.

Transmission Components:

Maintenance of transmission components is essential to ensure proper functioning. For manual transmissions, the gearbox, synchronizers and internal gears of the gearbox are subject to some wear over time. Maintaining an adequate level and quality of the clutch oil helps reduce wear and preserve performance, The life of the clutch disc can extend from 20,000 to 200,000 km all depends on many factors, in the presence of symptoms such as slipping, unpleasant odors or difficulty in changing gears it is advisable to replace it. During replacement, the condition of the flywheel and thrust bearing is also checked, with a possible replacement of these components. Manual transmission oil requires regular replacement, usually every 60,000 to 80,000 km or following the manufacturer's recommendations. In automatic transmissions, the torque converter is critical, the checks and oil change vary greatly according to the manufacturer's instructions. Sequential transmissions may require clutch replacement according to the same recommendations as manual transmissions, along with regular gearbox oil changes.

It is important to keep in mind that maintenance times and intervals may differ depending on the vehicle model and the manufacturer's recommendations. Investing in preventative maintenance is essential to prevent costly failures and ensure reliable and long-lasting transmission operation over time.

Importance of Technical Assistance and Periodic Overhauls

In this intricate world of gears and technologies, technical assistance and periodic inspections emerge as vigilant guardians of the longevity and efficiency of our vehicle. Turning your attention to a qualified professional is more than advice; It's a provident act that can mean the difference between a smooth ride and the risk of costly and dangerous inconveniences.

Service experts are the modern craftsmen of our automotive age, equipped with in-depth knowledge and specialized tools to scrutinize every nook and cranny of the vehicle. Periodic reviews play a crucial role in detecting potential issues before they escalate into more serious failures. From testing the braking system to advanced engine diagnostics, every component undergoes a rigorous examination to ensure its proper functioning.

Relying on qualified professionals not only helps maintain optimal vehicle performance, but also ensures the safety of occupants and other road users. Hidden problems can be identified and solved

promptly, avoiding dangerous situations and costly corrective actions in the future.

In addition, periodic inspections are an opportunity to adopt manufacturer-recommended improvements and updates, keeping the vehicle up to date with the latest technologies and safety regulations. Technical assistance is not just a necessity in response to obvious failures, but a proactive practice that invests in the overall health of the vehicle.

Finally, a detailed record of inspections and service interventions is a valuable document for the owner. Not only does it make it easier to manage deadlines and upcoming inspections, but it can also increase the resale value of the vehicle. Conscious buyers appreciate a car with a history of accurate maintenance, reflecting the owner's commitment to preserving the integrity of the vehicle over the years.

Technical assistance and periodic inspections are not simple formalities, but real investments in the most precious asset: the safety and reliability of our vehicle. With the expert eyes of service workers, we can keep our machine in top condition, ready to hit the roads with confidence and peak performance.

Useful Accessories to Always Keep with You in the Car

When venturing out on the road, it is advisable to be prepared for any unforeseen events. Some essential accessories to always keep with you in the car can make a difference in emergency situations or during prolonged journeys. Here are some tips to ensure a safer and more comfortable journey:

1. Portable Tyre Compressor: **A portable compressor is a valuable ally in the event of a flat tyre. Maintaining the right tire pressure is crucial for safety and fuel efficiency.**

2. Jump Starter: **A jump starter is useful for jump-starting the car in situations where the engine does not start due to a non-optimal state of the battery. It is a practical solution to avoid**

inconveniences or the use of cables that would inevitably require another car or an auxiliary battery.

3. **Snow Chains/Socks:** In areas subject to harsh winter conditions, snow chains are essential to ensure adequate traction on snowy or icy roads. Alternatively, snow socks are a viable option for improving traction in emergency situations, offering a faster and easier to install solution than traditional chains.

4. **First Aid Kit:** A well-stocked first aid kit is essential. It includes bandages, disinfectants, plasters and other items necessary for basic care in case of minor accidents.

5. **Battery-powered flashlight:** A battery-powered flashlight is crucial for lighting in nighttime emergency situations.

6. **Portable Charger (Power Bank):** A portable charger for electronic devices can be useful for keeping mobile phones and other devices charged during long trips.

7. **Hygienic bag:** A bag with basic hygiene items, such as paper towels, wet wipes, and waste bags, can make the travel experience more comfortable.

8. **Thermal blanket:** A compact thermal blanket can be invaluable in emergency situations or in the event of a prolonged vehicle stop.

9. **Multi-purpose tools:** A small set of multi-purpose tools, such as a multi-function screwdriver, can be useful for dealing with minor repairs.

10. **Fire extinguisher:** A fire extinguisher with a compact size and suitable for extinguishing small fires. Check its expiration date periodically and make sure it is in good condition.

Remember to adapt the accessories according to your personal needs and the type of trip you are facing. A well-equipped car can help ensure your safety and that of other road users.

Chapter 4: Types of Nutrition

The fourth chapter of our journey through automotive mechanics is dedicated to exploring revolutions in vehicle fuel systems. From traditional internal combustion engines to innovations in the electric space, we will look at the different types of power that are driving the evolution of the automotive industry. Through a comprehensive overview, we will analyze the characteristics, differences and challenges related to each option, taking a critical look at the roads that lead to more sustainable and efficient mobility.

Insight into the Various Types of Engines

In the wake of automotive evolution, our journey begins with an in-depth analysis of the different engines that have shaped the history of the automobile and continue to define its future. From the initial burst of petrol engines, which set the pace of the first automobiles, to the robustness and efficiency of diesel engines, which influenced the industry with their power and fuel economy, we will explore the historical roots that have nurtured the motoring variety.

Petrol engines, known for their versatility and immediate response, have gone through decades of technological advancements. From the beginnings with simple one- or two-cylinder engines, we have seen the evolution towards more complex engines with four, six or even eight cylinders, powering luxury cars and prestigious sports cars. Displacements have grown over the years, fueling the race for power and performance. The introduction of technologies such as direct injection and downsizing (the practice of using smaller engines with a reduced displacement, often accompanied by supercharging through the use of turbos or superchargers, to achieve performance similar to or superior to larger, more traditional engines) has further refined efficiency and emissions.

Diesel engines, on the other hand, have long been the bulwark of fuel efficiency. With high torque and higher thermal efficiency, they have found extensive use in commercial vehicles and touring cars. The evolution has led to the introduction of more refined diesel engines, with emission reduction technologies and a greater focus on sustainability.

Gas engines represent a category of propulsion that uses gas as a fuel source to generate power and power the engine. The main variants include natural gas (CNG), liquefied methane gas (LNG), biogas (produced through a process called anaerobic digestion), hydrogen and syngas engines. Using gas as a fuel offers cleaner emissions benefits than traditional fuels, with hydrogen considered a "clean" energy source since its combustion mainly produces water. However, the adoption of gas engines is influenced by the availability of cost-effective gas sources and the presence of suitable infrastructure for distribution.

Hybrid engines, a symbol of modern innovation, combine the best of both worlds, integrating internal combustion engines with electric units. These vehicles offer a smooth transition between electric and traditional modes, reducing emissions and increasing fuel efficiency. With mild hybrid and plug-in hybrid powertrain systems, the industry has embraced diversity to meet the needs of an increasingly aware clientele.

The next step leads us to the electric motor, a revolution in mobility. Battery-powered vehicles, powered solely by electricity, are gaining ground. The race to increase range, improve charging and reduce battery costs is at the heart of this era of transition. Electric car models stand out for their energy efficiency, quiet operation and reduced environmental impact.

Each type of engine is a page in automotive history, reflecting technological changes, consumer preferences, and the needs of our time. Through this in-depth study, we will focus on the distinctive features, from displacements to advanced technologies, offering a complete view of the engines that move the cars of the present and the future.

Difference Analysis Between Power Systems

In the broad landscape of automotive mechanics, detailed analysis of fuel systems becomes crucial to understanding the subtle nuances that characterize engines and their evolution over time. In this excursus, we will explore internal combustion engines, hybrid and all-electric vehicles in depth, analyzing their specific applications, historical transformations and technical impacts on performance and consumption.

Internal Combustion Engines - Petrol and Diesel

The history of the automobile is intimately linked to gasoline-powered internal combustion engines. From the first models at the beginning of the century to modern engines, this technology has undergone a remarkable evolution. Initially characterized by structural simplicity, petrol engines have seen progressive complexification to improve efficiency, increase power and reduce emissions. The introduction of direct injection was one of the pivotal moments, allowing for more precise control of fuel delivery and an increase in thermal efficiency.

In parallel, diesel engines have made an appearance, becoming the preferred choice for heavy-duty vehicles and SUVs. The key feature of diesel engines is fuel efficiency, thanks to the more efficient combustion and higher energy density of diesel fuel (35 megajoules per liter (MJ/L) or 45-48 megajoules per kilogram (MJ/kg)). Over the years, diesel engines have undergone major technological innovations to reduce harmful emissions, including particulate filters and SCR (Selective Catalytic Reduction) systems.

The evolution of both types of engines has led to a race for efficiency, with technologies such as downsizing (reducing displacements to improve thermal efficiency) and the implementation of turbocharging systems. The continuous search for a balance between power and sustainability has made these engines more ecological and performing over time.

The Battle of Ibridi

The entry of hybrid vehicles has redefined the very concept of automotive propulsion. These vehicles combine an internal combustion engine with an electric motor, taking advantage of the best of both worlds. The mild hybrid variants use the electric motor mainly in low-speed driving situations, thus reducing fuel consumption in the city. On the other hand, plug-in hybrid models allow for all-electric driving for longer distances, with the internal combustion engine kicking in when needed.

The specific application of hybrid vehicles makes them particularly suitable for a wide range of driving scenarios. Mild hybrid models find their way into urban areas, where stop-and-go traffic lends itself to the use of the electric motor. Plug-in hybrid vehicles, on the other hand, are ideal for those looking for a more sustainable solution for long distances, being able to rely on the electricity grid for overnight charging.

Electric Vehicles

The electric revolution has transformed the concept of mobility. All-electric vehicles (EVs) are powered solely by batteries, eliminating the internal combustion engine entirely. The evolution of batteries has been at the heart of this revolution, with constant research to increase range, reduce charging times and keep costs down.

EVs have mainly established themselves in urban settings, thanks to their quiet driving, low emissions and growing charging infrastructure in cities. Growing consumer interest and government incentives have accelerated the production of EV models, with many automakers announcing ambitious strategies for the future, aiming to electrify their entire range.

Evolutionary Trends and Technical Impact

Evolutionary trends in fuel systems continue to drive innovation in the automotive industry. The miniaturisation of internal combustion engines has been a

response to the growing demand for efficiency and compactness. Smaller engines, however, have not sacrificed power, thanks to turbocharging and the optimisation of electronic management technologies. Advancement in lightweight materials, such as aluminum alloys and carbon fiber, has helped to make vehicles lighter, improving fuel efficiency and overall performance. The search for sustainable solutions has prompted many manufacturers to explore ecological materials and production processes with low environmental impact.

The adoption of advanced technologies, such as autonomous driving and driver assistance systems, has integrated with the evolution of power systems, redefining the driving experience in terms of safety and comfort. The interconnection of these technological developments has led to synergistic innovation, where vehicles increasingly become an integral part of an ever-evolving digital ecosystem. The in-depth analysis of the differences between the fuel systems goes beyond mere technical description, as it reveals a dynamic narrative of changes in the automotive landscape. This narrative is intertwined with changing consumer needs, environmental pressures, and the race for innovation, driving the industry towards more sustainable and advanced mobility.

Advantages and Disadvantages of Automotive Powertrain Technologies

This section is dedicated to the "Advantages and Disadvantages of the various propulsion technologies" and aims to examine in detail the different engine options, focusing on diesel and gasoline internal combustion engines, as well as electric and hybrid technologies.

Internal Combustion Engines, Petrol and Diesel

Advantages:

- **Power and Acceleration:** Internal combustion engines offer remarkable power, ensuring prompt acceleration and dynamic driving, ideal for performance enthusiasts.
- **Range and Refuelling:** Petrol and diesel cars have a long range and can be refuelled easily at a large network of petrol stations, making them ideal for long journeys.
- **Traditional Driving Experience:** For those who love classic driving, internal combustion engines offer a unique sound and tactile experience, establishing an emotional bond between the driver and the vehicle.

- ➤ Generally lower upfront costs: Compared to more advanced alternatives, internal combustion vehicles often have lower upfront costs, making them accessible to a wider audience.

Detriments:

- ➤ Pollutant emissions: The main criticism is represented by harmful emissions, contributing to air pollution and the greenhouse effect.
- ➤ Dependence on Fossil Fuels: The need for fossil fuels implies a dependence on non-renewable resources, with consequent impacts on the environment.
- ➤ Variable Operating Costs: Fuel costs are subject to fluctuations, influenced by the dynamics of the global oil market.
- ➤ More Frequent Maintenance: Internal combustion engines often require regular maintenance and replacement of components, increasing costs in the long run.

The Battle of Ibridi

Advantages:

➢ Fuel Efficiency: The integration of internal combustion engines and electric motors offers high fuel efficiency, especially in urban driving conditions.

➢ Energy Recovery: Hybrid systems often recover energy during braking, improving the overall efficiency of the vehicle.

➢ Reduced Dependence on Fossil Fuels: The use of the electric motor partially reduces the dependence on fossil fuels when driving at low speeds or vehicles with the possibility of external charging.

➢ Mixed Driving Experience: Hybrid vehicles offer a driving experience that combines traditional and advanced elements, adapting to a wide range of driver preferences.

Detriments:

➢ High upfront costs: Hybrid vehicles may have higher upfront costs than traditional internal combustion vehicles.

- **Technological Complexity:** The presence of two propulsion systems increases the complexity of the vehicle, potentially leading to higher maintenance costs.

- **Fuel Dependency:** Despite fuel efficiency, hybrid models continue to depend on fossil fuels for high speeds or hybrids that cannot have external charging.

- **Material Resources for Batteries:** The production of batteries uses materials such as lithium, cobalt, nickel and sometimes rare earths such as cerium, neodymium, praseodymium, europium, terbium, dysprosium, gadolinium, yttrium and others play a key role in various technological applications, generating potential environmental impacts during the extraction of these elements.

Electric Vehicles

Advantages:

- **Zero Local Emissions:** EVs emit no local emissions while driving, reducing air pollution in urban areas.
- **Reduced Operating Costs:** EVs are generally cheaper to operate, with lower energy costs than internal combustion vehicles.
- **Simplified Maintenance:** With fewer moving parts than traditional motors, EVs often require less maintenance.
- **Quiet and Comfortable Ride:** The absence of engine noise contributes to a quieter and more comfortable driving experience.

Detriments:

- **Limited Range and Charging Time:** The range of EVs can be limited, and charging times longer than traditional refuelling can be an annoying inconvenience.
- **Charging Infrastructure:** The availability of charging stations can be limited, especially in less developed areas.
- **High Upfront Costs:** EVs often have higher upfront costs than traditional vehicles.

- **Environmental Challenges in Battery Production:** Battery production, as mentioned above for hybrid vehicles, comes with environmental challenges related to the extraction of materials and the management of waste from spent batteries.

This detailed examination highlights that choosing between these technologies requires a thoughtful assessment of the advantages and disadvantages, considering individual needs, driving preferences and environmental awareness. With continued technological development, many of the current challenges could be overcome, contributing to a smoother transition to sustainable mobility.

Chapter 5: Motorsport Technologies, Analysis of Different Existing Solutions

In this chapter, we will dive into engine technologies specifically, taking a close look at the various powertrain solutions that power the automotive industry. From internal combustion engines to hybrid powertrains and all-electric vehicles, we will have the opportunity to explore the innovations that define the way we get around. Lifting the veil on the complexity of these technologies, we will discover the beating heart of the different engines, intertwining stories of power, efficiency and sustainability. An in-depth analysis that will take us through the pages of the evolution of automotive propulsion and the challenges faced to shape the future of mobility.

Internal Combustion Engines: Petrol and Diesel

Petrol and diesel engines, while belonging to the internal combustion engine category, have fundamental differences in the way they generate power. One of the key elements that distinguishes them is the type of fuel used. Petrol engines operate mainly on petrol (also called petrol), while diesel engines rely on diesel as their primary energy source.

Differences between Petrol and Diesel Engines

1. Combustion and Ignition:

Petrol Engines: The distinctive feature of petrol engines is the combustion process that takes place thanks to a spark generated by an ignition plug. This method, known as the Otto cycle, follows four basic steps: intake, compression, combustion and exhaust. During the intake phase, the air-fuel mixture is introduced into the engine cylinder. Next, the piston moves upwards in the compression phase, compressing the mixture. The explosion occurs when the spark plug generates a spark, triggering combustion and pushing the piston down. Finally, the exhaust phase expels the exhaust gases generated during combustion.

Diesel Engines: Unlike gasoline engines, diesel engines use a different approach to combustion. The spark generated by the spark plug is replaced by the compression of the air in the cylinder. This is part of the Diesel cycle, which also follows the four phases of intake, compression, combustion and exhaust. During the compression phase, the air is compressed significantly, generating heat. At some point, diesel fuel is injected directly into the compressed air, and the high temperature triggered by compression ignites the fuel, initiating combustion. The piston is then pushed down, completing the cycle. This feature makes diesel engines more fuel-efficient than gasoline engines.

2. Yield and Performance:

Gasoline Engines: Gasoline engines are often preferred for high-speed performance. Gasoline burns faster than diesel, providing quick throttle response. This feature makes them ideal for sports vehicles or situations where immediate response is needed when accelerating. They tend to operate at higher rpm, reaching maximum power at higher speeds. However, petrol engines may be less fuel-efficient than diesels, especially when driving under heavy load or in the city, where speeds can vary greatly.

Diesel Engines: Diesel engines are known for their fuel efficiency and ability to produce high torque at low rpm. This makes them ideal for vehicles that require more relaxed driving, such as highways or long distances. The high torque allows for a more responsive response at low speeds, which is especially useful in towing or heavy load situations. While diesel engines are generally more fuel-efficient, they may be less suitable for situations where very high speeds are required. In addition, diesel engines tend to run at lower rpm than petrol engines. The choice between a petrol and diesel engine depends on the user's specific needs, such as the type of driving, frequency of use and personal preference for performance or fuel efficiency.

To optimize the efficiency of diesel and gasoline engines, the use of specific fuels and targeted additives can be considered. Premium diesel (Diesel+ or Blu Diesel), enriched with detergents and stabilizers, contributes to cleaner combustion, while biodiesel, derived from renewable sources, offers an environmentally friendly option. In the gasoline engine category, premium gasoline (100 octane) with detergent additives improves the cleanliness of engine components, and greener formulations reduce pollutant emissions. In summary, premium fuels can offer benefits such as improved performance and extended vehicle life when used in accordance with the manufacturer's recommendations

Technological Evolution

In addition to the key differences, both types of engines have undergone significant technological advancements to address environmental challenges, improve fuel efficiency, and provide optimal performance.

1. Direct Fuel Injection:

> Both types of engines have adopted direct fuel injection systems, allowing for more precise control of combustion and improving engine efficiency. Its implementation brings numerous benefits, including:
>
> ❖ Precision Mixing: **Fuel is atomized directly into the combustion chamber, improving the accuracy and distribution of air-fuel mixing.**

- ❖ Combustion Control: **The technology allows precise control over the quantity and time of injection, optimizing combustion and reducing energy losses.**

- ❖ Thermal Efficiency: **Accurate fuel distribution leads to more complete combustion, increasing the thermal efficiency of the engine and improving mileage.**

- ❖ Improved Performance: **Thanks to precise injection management, optimized performance is achieved in terms of power, torque and throttle response.**

- ❖ Reduced Emissions: **More efficient combustion helps reduce pollutant emissions while adhering to the most stringent environmental regulations.**

- ❖ Dynamic Adaptability: **The technology dynamically adapts to different driving conditions, adjusting the injection in real time to meet the driver's needs and vehicle conditions.**

- ❖ Multiphase systems**: Some engines implement multiphase systems, allowing the quantity and pressure of the fuel injected to vary in different phases of operation.**

Direct Fuel Injection represents a significant step towards more efficient, high-performance and environmentally friendly engines, highlighting how technological innovation continues to drive the evolution of the automotive sector.

2. The use of superchargers such as turbochargers:

Many applications, both petrol and diesel, use turbocharging, it is a revolutionary engineering technology that has redefined the characteristics of internal combustion engines, bringing significant improvements in performance and efficiency. This innovation, increasingly integrated into modern engines, has transformed the way vehicles generate power.

- ❖ Working Principle: The turbocharger uses the energy of the engine's exhaust gases to power a turbine connected to a compressor. This compressor increases the pressure of the air sucked into the engine, allowing for more efficient combustion.

- ❖ Increased Power and Torque: Higher air pressure allows the engine to burn more fuel, generating a significant increase in engine power and torque without the need to enlarge the engine's size.

- ❖ Quick Response: The turbocharger reduces the so-called "turbo-lag", i.e. the delay between the request for acceleration and the actual response of the engine. This results in faster throttle response and more dynamic driving.

- ❖ Fuel Efficiency: A turbocharged engine can achieve higher fuel efficiency. The higher air pressure during combustion contributes to a more efficient use of the intrinsic energy present in the fuel, improving the overall yield of the process.

- ❖ Reduced Emissions: Improved combustion efficiency thanks to the turbocharger contributes to a reduction in pollutant emissions, supporting efforts to ensure cleaner vehicles that comply with environmental regulations.

- ❖ Varied Applications: The turbocharger is used in a wide range of engines, from petrol to diesel. This versatility proves its effectiveness in different contexts and types of vehicles.

- ❖ Future trends: New technologies, such as variable geometry turbochargers, are emerging to further improve the efficiency and adaptability of the turbocharger to various driving conditions.

The turbocharger, with its ability to increase power, improve efficiency and reduce emissions, continues to remain a key element in modern engine technologies, demonstrating how constant innovation drives the evolution of the automotive industry.

3. Cylinder Deactivation Systems:

The cylinder deactivation system operates by temporarily disabling a part of the engine cylinders when the power required is less than the maximum available. For example, in situations when driving at a constant speed or at low engine speeds, where the power required is limited, the system can deactivate one or more cylinders.

- ❖ **Condition Monitoring:** An electronic control system constantly monitors driving conditions, evaluating factors such as vehicle speed, engine load, throttle pressure, and other relevant parameters.
- ❖ **Execution and Deactivation:** When the vehicle is operating in driving conditions that require lower engine power, the system may decide to deactivate some cylinders. This is done by stopping fuel injection and turning off ignition in selected cylinders.

- ❖ **Reduced Cylinder Operation:** During deactivation, cylinders not involved in the combustion process operate in "free" mode, without consuming fuel and without contributing to power generation. This way, the engine temporarily runs with fewer active cylinders.
- ❖ **Automatic Reactivation:** When the required power increases, the system automatically reactivates previously deactivated cylinders, allowing the engine to operate with the maximum power available.

Advantages:

- ➤ **Reduced Fuel Consumption:** Deactivating cylinders results in a significant reduction in engine load in light driving conditions, helping to improve fuel efficiency.
- ➤ **Reduced Emissions**: Decreasing the engine load when deactivating cylinders can also lead to a reduction in pollutant emissions.
- ➤ **Improved Energy Efficiency:** The system allows the engine to dynamically adapt to power demands, improving the overall energy efficiency of the vehicle.

Detriments:

- ➤ Mechanical and Electronic Complexity: **System** implementation requires increased mechanical and electronic complexity in the engine, potentially increasing manufacturing and maintenance costs.
- ➤ Possible Impact on Engine Life: **Some** critics argue that frequent deactivation and reactivation of cylinders could have an impact on engine life, even as manufacturers try to mitigate this problem with robust designs.
- ➤ Driving Feel: **In** some cases, drivers may feel a slight difference in throttle response or vibration when transitioning between cylinder deactivation and activation modes.

In general, despite the potential drawbacks, cylinder deactivation systems are considered an effective technology for improving the efficiency of motor vehicles, especially in light driving situations or at constant speeds. The technology has been adopted on several vehicles to meet the growing demands for fuel efficiency and emission reduction. The continuous evolution of these technologies reflects the automotive industry's commitment to balancing performance, efficiency, and sustainability to meet the ever-increasing needs of consumers and address environmental challenges.

Hybrid Engines: A Journey Through Parallel Configurations, Series and Other Technologies

Interest in hybrid dates to the early decades of the twentieth century, with some prototypes combining internal combustion engines with electric motors. However, it is in recent decades that hybrid technology has experienced exponential growth, driven by growing environmental concerns and the search for more fuel-efficient solutions.

Hybrid engines combine a combustion engine with an electric motor to improve vehicle efficiency and reduce emissions. There are different configurations of hybrid engines, each with its own characteristics. Here are some of the main types of hybrid engines:

Parallel Hybrids

Operation:

In parallel hybrid vehicles, the system architecture allows both the internal combustion engine and the electric motor to directly power the vehicle's wheels. The two motors can work together, providing combined power. The specific operation of a parallel hybrid may vary slightly between models, but it generally follows a few key principles:

1. **Hybrid Mode:** Under normal driving conditions, the vehicle can operate in hybrid mode, with both engines running. During acceleration or when more power is required, the system can activate both the internal combustion engine and the electric motor at the same time.

2. **Electric Only Mode:** In some situations, especially at low speeds or when driving lightly, the vehicle can switch to all-electric mode, powered by the battery and electric motor alone. This reduces or eliminates the use of the internal combustion engine, contributing to greater fuel efficiency and emissions.

3. **Deceleration and Energy Recovery:** During deceleration or braking, the system can take advantage of regenerative braking to convert kinetic energy into electrical energy. This energy can be used to recharge the battery, helping to keep enough charge for future accelerations.

Example: Toyota Prius

The Toyota Prius is one of the most well-known examples of a parallel hybrid vehicle. Its hybrid architecture allows the petrol engine and electric motor to work synergistically. When driving at low speeds or when modest power is required, the Prius can run solely on the electric motor. When the demand for power increases, the petrol engine also comes into play, providing the necessary extra power.

This system offers greater fuel efficiency than a traditional vehicle, contributing to more sustainable driving. The parallel hybrid approach represented by the Prius has been instrumental in popularizing hybrid technology and has influenced the design of numerous other hybrid models on the market.

Hybrids Series

Operation:

In serial hybrid vehicles, the system architecture stipulates that the internal combustion engine operates exclusively to power an electric generator. This generator, in turn, produces the energy necessary to power both the battery and directly the electric motor, which is solely responsible for the traction of the wheels.

This approach creates a unidirectional flow of power, with the heat engine acting as a generator for the electric motor.

1. **Generator Mode:** While driving, the internal combustion engine operates in generator mode, converting the energy from the combustion of fuel into electrical energy. This energy is then stored in the batteries or used directly to power the electric motor.

2. **Electric Motor for Traction:** The electric motor is the key element for the traction of the vehicle. It is responsible for converting the electrical energy supplied by the generator into the movement of the wheels.

3. **Battery Charging:** During deceleration or braking, the electric motor can also act as a generator, converting kinetic energy into electricity to recharge the batteries. This process helps to maintain a reserve of energy to be used when more power is required.

Example: BMW i3

The BMW i3 adopts a series hybrid architecture. In this configuration, the internal combustion engine acts solely as a generator to produce electricity.

This energy is then used to power the electric motor, which provides traction to the wheels. This approach allows the BMW i3 to reap the benefits of an electric motor's efficiency while maintaining the flexibility offered by the internal combustion engine, ideal for situations where electric charging may not be readily available. The series hybrid configuration can improve fuel efficiency and reduce emissions, especially in urban driving contexts and in heavy traffic situations. In practice, the internal combustion engine works autonomously at optimum operating speeds.

Plug-in hybrids (PHEV)

Operation:

Plug-in hybrid vehicles (PHEVs) are characterized by the ability to operate both in all-electric mode and with the combustion engine. The main distinction from traditional hybrids is the ability to recharge the batteries from an external source, such as a household power outlet or a public charging station. The operation of a PHEV can be divided into the following key points:

1. Electric Mode: When the batteries are charged, the vehicle can operate in all-electric mode, using only the electric motor for traction. This mode is ideal for urban driving at low speeds or for covering short distances without using the internal combustion engine.

2. Hybrid Mode: When the battery charge decreases or when more power is required, the vehicle can switch to hybrid mode. In this mode, the internal combustion engine kicks in to provide additional power to the electric motor, allowing it to cover longer distances and maintain high performance.

3. External Charging: A distinctive feature of PHEVs is the ability to recharge the batteries by connecting the vehicle to an external source of electrical power. This allows for an increase in range in all-electric mode, reducing dependence on the internal combustion engine.

Example: Chevrolet Volt, Mitsubishi Outlander PHEV

The Chevrolet Volt and Mitsubishi Outlander PHEV are examples of plug-in hybrid vehicles. The Chevrolet Volt is known for its ability to run in all-electric mode for a significant distance before activating the combustion engine. The Mitsubishi Outlander

PHEV, on the other hand, is a plug-in hybrid SUV that offers the flexibility of an electric motor for urban driving and the ability to use the internal combustion engine for longer journeys. These plug-in hybrid vehicles offer a compromise between all-electric driving and extended range, making them suitable for a variety of driving needs and helping to reduce environmental impact as usual.

Mild Hybrid

Operation:

Mild hybrid vehicles are characterized by a system in which the electric motor acts as a support unit for the internal combustion engine but does not have the ability to move the vehicle on its own. These systems are designed to improve fuel efficiency and reduce emissions without the capability of all-electric driving. The operation of a mild hybrid system can be divided into the following key aspects:

1. Assistance during Acceleration: During acceleration phases, the electric motor provides additional support to the internal combustion engine. This can reduce the strain required by the internal combustion engine, improving fuel efficiency and providing faster throttle response.

2. Energy Recovery: During deceleration and braking, the electric motor acts as an electric generator, converting kinetic energy into electricity. This energy is then used to recharge the battery, helping to maintain a reserve of energy for future accelerations.

3. Stop & Start: Some mild hybrid systems may implement an automatic stop & start function, which shuts off the internal combustion engine when the vehicle is stationary and restarts it when the accelerator is pressed. This helps to save fuel during breaks and at traffic lights.

Example: Honda Accord Hybrid, Kia Niro Hybrid

The Honda Accord Hybrid and Kia Niro Hybrid are examples of mild hybrid vehicles. In both cases, the electric motor assists the internal combustion engine to improve fuel efficiency and reduce emissions. These vehicles are ideal for those looking for improved performance and efficiency without the need for all-electric driving. Mild hybrid systems are a step towards more sustainable solutions in the automotive industry, helping to reduce environmental impact and improve fuel efficiency.

Diesel hybrids

Operation:

Diesel hybrid vehicles combine hybrid technology with the efficiency of diesel engines to maximize fuel efficiency. These vehicles use the diesel engine in conjunction with an electric motor to achieve greater fuel efficiency and reduced emissions. The operation of a diesel hybrid can be divided into the following key points:

1. Diesel Engine: **The diesel engine, known for its fuel efficiency and high torque production at low rpm, plays a vital role in the propulsion of the vehicle. It provides consistent and efficient power while riding.**

2. Electric Motor: **The electric motor provides assistance to the diesel engine during acceleration and in situations where more power is required. It can also operate on its own in all-electric mode at low speeds or in light riding conditions.**

3. Battery Charging: **During deceleration or braking, the electric motor can act as a generator, converting kinetic energy into electricity to recharge the batteries. This process helps to maintain a reserve of electrical energy to be used when electric motor support is needed.**

Example: Peugeot 3008 Hybrid4

The Peugeot 3008 Hybrid4 is an example of a diesel hybrid vehicle. This model uses the combination of a diesel engine with an electric motor. The hybrid system makes it possible to use the power of the diesel engine when a more robust drive is required and to take advantage of the electric motor for a quiet and low-emission drive in light traffic situations.

Through this combination, diesel hybrids seek to offer the best of both technologies, the fuel efficiency of diesel engines and the energy efficiency of electric motors. These vehicles represent an attempt to balance performance, efficiency and reduced emissions in road mobility.

Compressed Air Hybrids

Operation:

Compressed air hybrid vehicles represent an innovative experiment in which internal combustion engines are powered primarily by energy generated through compressed air. The dynamics of a compressed air hybrid are divided into the following fundamental aspects:

1. **Compressed Air Power Generation:** During the deceleration and braking phases, kinetic energy is converted into compressed air through a compression process. This compressed air is then stored in special tanks.

2. **Internal combustion engine:** The internal combustion engine of a compressed air hybrid vehicle draws energy primarily from air compression, using this process to generate mechanical power through optimized fuel combustion. This dynamic is characterized by an increase in temperature and pressure during air compression, followed by the combustion of fuel to transform the energy of the compressed air into piston movement and, finally, into mechanical power.

3. **Electric Motor:** The electric motor can act as an additional support, especially during acceleration phases or when more power is required. It can also act as a generator during deceleration, helping to recharge the electric battery or maintain air compression.

Example: R&D Prototypes

Currently, compressed air hybrids are mainly the subject of research and development. Some companies and institutions are experimenting with concept prototypes that seek to harness compressed air as the primary energy source for internal combustion engines.

It is important to note that now, this technology is still being studied and is not widely used in the mass automotive market. However, it represents an interesting field of research to explore new energy solutions and reduce the environmental impact of internal combustion vehicles.

Hydrogen Engines: Hydrogen Engines Exploration

Hydrogen engines represent one of the most fascinating solutions in the race towards sustainable mobility. Their peculiarity lies in the fact that hydrogen, a light and abundant gas, serves as the main fuel. This approach offers significant benefits in terms of zero emissions and reduced environmental impact.

How it works: Hydrogen engines use a technology known as fuel cells. In this process, hydrogen reacts with oxygen in the air inside a fuel cell, generating electricity, water, and heat as byproducts. This electricity is then used to power the vehicle's electric motor, providing the traction needed for driving.

Advantages:

1. **Zero Local Emissions:** The reaction between hydrogen and oxygen produces only water and heat as by-products, eliminating harmful emissions on site and helping to improve air quality in urban areas.

2. **High Energy Efficiency:** The direct use of electricity generated by the fuel cell means more efficient energy conversion than traditional internal combustion engines.

3. **Extended Range:** Hydrogen vehicles can offer a range comparable to conventional gasoline, diesel, or gas-powered vehicles, solving one of the key challenges of electrification.

Challenges and Obstacles:

1. **Hydrogen production:** Hydrogen production currently requires a considerable amount of energy, and hydrogen sources must be developed sustainably to maximize environmental benefits.

2. **Charging Infrastructure:** The availability of hydrogen refuelling stations is limited compared to traditional or

electric charging stations, making the charging infrastructure an obstacle to overcome for large-scale deployment.

3. **High upfront costs:** Currently, hydrogen vehicles can be expensive to produce and purchase, but technological developments and growing demand could reduce these costs over time.

Hydrogen engines emerge as a promising prospect compared to traditional sources of energy. Although they are still in development and must overcome some challenges, their adoption could outline a new era of sustainable and zero-emission mobility. It should be noted, however, that vehicles that use hydrogen for internal combustion do not fit into this positive picture, being generally less efficient than vehicles equipped with fuel cells. The latter, thanks to their greater efficiency in converting hydrogen into electricity, are presented as a more promising solution from an energy and environmental point of view.

Fuel Cell Operation

The fuel cell, especially proton exchange polymer membrane (PEMFC) type fuels, is a critical component in hydrogen vehicles. Let's now see in detail how this technology works:

1. Introduction of Hydrogen: **The process begins with the introduction of hydrogen (H2) into the combustion cell. The hydrogen is stored in the vehicle's tanks and is used to power the cells when electricity production is needed for traction.**

2. Anode and Catalyst: **Inside the combustion cell, hydrogen meets the anode, which is coated with a catalyst. In the presence of this catalyst, hydrogen dissociates into its positive (protons) and negative (electrons) ions. The chemical reaction at the surface of the anode can be represented as follows:** $H2 \rightarrow (2H+) + (2e-)$

3. Proton Exchange Membrane (PEM): **The proton exchange membrane is the heart of the combustion cell. This membrane is permeable to protons (H+) but not to electrons (e-). The protons generated in the anode pass through the membrane, while the electrons travel through an external circuit.**

4. **External Electrical Circuit:** Electrons, being negatively charged, travel through an external circuit, generating electric current in the process. This current is available to power a small battery or directly to the vehicle's electric motor.

5. **Catalyst and Cathode Reaction:** The electrons, after traveling through the external circuit, reach the cathode, which is coated with another catalyst. At the cathode, H+ ions from the membrane and electrons combine with oxygen in the air to form water:

$$(2H+) + (1/2*O2) + (2e-) \rightarrow H2O$$

This reaction is exothermic, generating heat as a byproduct.

6. **Water Production:** The water produced during the reaction at the cathode is expelled from the cells in the form of water vapor. This represents the only by-product of the electricity generation process, giving this technology a characteristic of zero local emissions.

The overall efficiency of the combustion cell is determined by the direct conversion of hydrogen into electricity, without the need for intermediate mechanical converters as in traditional internal combustion engines. While PEMFCs are commonly used in vehicles, there are also other types of fuel cells, each with its own specific characteristics and applications.

Advanced Electric Motors: Batteries, Motors and Emerging Technologies

In the dynamic world of sustainable mobility, electric vehicles are experiencing an accelerated technological transformation. From state-of-the-art electric motors to high-capacity batteries, this section explores the all-important change behind the electric revolution. With increasingly efficient and safe energy storage solutions, we are witnessing a redefinition of the concept of propulsion. These components not only guide the choice for a sustainable option, but also symbolize technological excellence.

Permanent Magnet Motors: High Efficiency and Performance

Permanent magnet (synchronous) motors are an innovative and increasingly popular propulsion technology in the automotive industry. These motors harness permanent magnet energy to generate the driving force needed to move the vehicle. Their growing popularity is due to their energy efficiency, improved performance.

Operation:

1. **Stator and Rotor:** The motor consists of a stator (fixed part) and a rotor (moving part). The stator consists of wire windings arranged around a ferromagnetic core. These windings are powered by direct electric current, generating a fixed magnetic field

2. **Rotor magnetization:** The rotor is made up of permanent magnets, usually made of materials such as neodymium, which generate a constant magnetic field.

3. **Torque Generation:** The interaction between the magnetic field of the stator and the permanent magnets of the rotor generates torque, inducing the rotary movement of the rotor

4. **Electronic Control:** The electronic control system regulates the current supplied to the stator to control the speed and torque of the motor.

Advantages:

1. **High energy efficiency:** Motors are known for their high efficiency, which results from the constant presence of the magnetic field generated by permanent magnets

2. **Improved performance:** These engines are capable of providing remarkable performance in terms of acceleration and top speed, making them ideal for sports vehicles

3. **Quick Response:** The permanent magnet configuration allows for a quick response to changes in speed, contributing to a more responsive ride.

4. **Low maintenance:** As they contain no moving wear parts such as brushes and slip rings, they require minimal maintenance, reducing long-term running costs

Detriments:

1. **Material Costs:** The use of high-quality permanent magnets can result in higher costs in the production of the motor itself.

2. **Vulnerability to high temperatures:** In certain operating situations, such as in the presence of very high temperatures, they may experience a decrease in performance

3. **Construction Complexity:** Some permanent magnet motors can be more complex in design and construction, which further impacts production costs

4. **Difficulty in recycling:** Recycling magnetic materials can be problematic and expensive, contributing to the environmental impacts of e-waste disposal

In summary, permanent magnet motors are widely appreciated for the efficiency and high performance they offer, making them particularly suitable for advanced automotive applications. Ongoing research aims to mitigate any drawbacks and further improve the performance of this technology.

Induction Motors: Simplicity and Reliability in Tesla Model S

Induction motors, also known as asynchronous motors, are a technology that is widely used in the automotive industry and many other industrial applications. These motors are appreciated for their reliability, simplicity of construction and energy efficiency, these motors must work with alternating current so the vehicle must have an inverter that converts direct current from the batteries into alternating current.

Operation:

1. **Stators and Rotors:** The motor is composed of a rotor (moving part) and a stator (fixed part), the stator has the same characteristics as previously described, while the rotor is composed of conducting bars wound around a ferromagnetic core

2. **Rotating Magnetic Field:** When current is applied to the stator, a rotating magnetic field is generated. This rotating magnetic field induces an electric current in the rotor through the principle of electromagnetic induction.

3. **Driving force:** The interaction between the rotating magnetic field of the stator and the current induced in the rotor generates a driving force, causing the rotor to spin. This motion is then transmitted to the vehicle's wheels to generate motion.

4. **Electronic Control:** The electronic control system regulates the current supplied to the stator to control the speed and torque of the motor. This adjustment is crucial to ensure optimal performance in different driving conditions.

Advantages:

1. **Simplicity of Construction:** Induction motors are relatively simple to build, with fewer mechanical parts than some other types of motors, reducing the risk of failure.

2. **Low Maintenance:** Due to their simplified structure, induction motors generally require less maintenance than more complex motors, resulting in lower operating costs.

3. **Reliability:** Their simplified structure and lack of mechanical wear parts can contribute to greater reliability in the long term.

Detriments:

1. **Low specific power:** At low rpm, induction motors can be less efficient than some types of permanent magnet motors, which can affect vehicle performance, especially under high load conditions.

2. **Complexity of High-Speed Control:** Controlling speed at very high rpm may require more complex control systems to maintain engine efficiency and stability.

3. Size and Weight: Although less complex than some motors, induction motors can be bulkier and heavier for the same power output.

4. Sensitivity to voltage changes: They can be sensitive to voltage changes, which can affect their performance and reliability.

Induction motors offer a combination of simple construction and reliability, making them a popular choice for electric vehicles, such as the Tesla Model S. The continued evolution of technology could help overcome some of the associated drawbacks, further improving performance and efficiency.

Variable Reluctance Synchronous Motors: Efficiency in Jaguar I-PACE Vehicles

These motors are distinguished by their ability to vary their reluctance, i.e. resistance to magnetic field flow, in order to optimize performance under different load and speed conditions.

Operation:

Variable reluctance synchronous motors are designed to maximise efficiency in electric vehicles such as the Jaguar I-PACE. Their operation can be divided into the following steps:

1. **Variable Reluctance:** Reluctance is the tendency of a material to oppose the flow of the magnetic field. In variable reluctance synchronous motors, the design allows the reluctance, i.e. resistance to magnetic flux, to be adjusted dynamically.

2. **Advanced Controls:** The motor control system adjusts the rotor reluctance in real time. This allows to optimize the interaction between the magnetic field generated by the stator and the rotor, improving the efficiency of the motor in different driving conditions.

3. **Minimization of Losses:** By adjusting reluctance, energy losses caused by eddy currents and heating are minimized, improving the overall efficiency of the motor.

Advantages:

1. **High Efficiency:** Variable reluctance synchronous motors can offer remarkable efficiency due to the ability to dynamically

adapt to driving needs. Reluctance regulation helps to minimise energy losses.

2. Good Performance at All Rpm: These engines are designed to perform well in a wide range of driving conditions, ensuring efficient operation at both low and high rpm.

3. Simplicity of Construction: Compared to certain more complex motors, variable reluctance synchronous motors can perform well with a relatively simple design, contributing to lower production costs.

Detriments:

1. Initial Costs: Despite their simplicity of construction, implementing variable reluctance synchronous motors can result in higher upfront costs than certain types of more conventional motors.

2. Control Complexity: Dynamic reluctance adjustment requires advanced systems, which can result in increased complexity and associated costs.

3. Material Sensitivity: The optimal performance of these motors can be influenced by the quality and characteristics of the materials used in their construction.

In short, variable reluctance synchronous motors represent a compromise between efficiency, performance, and construction complexity. Continued research and development could help overcome some of the associated disadvantages and make this technology increasingly competitive in the EV landscape.

Linear Motors: Advanced Propulsion in High-Speed Maglev Trains

In this section, we explore Maglev technology, which was originally used in trains, and its possible application in the automotive industry. Focusing on aspects such as magnetic levitation and non-contact propulsion, we examine how this technology could affect wireless suspension, propulsion and charging systems in cars. The aim is to understand how Maglev innovation, designed for high-speed trains, could transform the automotive sector, leading to improvements in energy efficiency, design and overall driving experience, with a focus on the challenges and opportunities that come with it.

Operation:

Linear motors are essential in Maglev (magnetic levitation) high-speed trains. These trains use the principle of electrodynamics to generate the force necessary to lift and move the train without physical contact with the rails. The operation can be divided into three main phases:

1. Magnetic Levitation: **The train is equipped with permanent magnets on the underside, while the track has magnetic coils. As the train approaches, a magnetic force of repulsion is created, lifting the train above the track without any physical contact.**

2. Linear Propulsion: **Once levitated, the train is propelled forward by linear motors positioned along the track. These motors, which can be based on different technologies, generate a rotating or oscillating magnetic field that interacts with the magnets on the train, propelling it forward along the route.**

3. Speed Control and Direction: **The speed of the train is controlled by varying the frequency of the magnetic field generated by the linear motors. In addition, the direction of movement can be controlled by varying the polarity of the magnetic field.**

Advantages:

1. **High Speeds:** Maglev trains can reach very high speeds thanks to magnetic levitation and linear propulsion, significantly reducing travel times compared to conventional trains.

2. **Quietness:** Because there are no mechanical parts in contact, Maglev trains are known for their quiet operation, contributing to a better travel experience.

3. **Reduced Maintenance:** The lack of physical contact between the train and the track reduces wear and tear, contributing to lower maintenance costs than traditional trains.

Detriments:

1. **High Upfront Costs:** Building Maglev infrastructure, including linear motors and magnetic levitation systems, incurs significant upfront costs.

2. **Technological complexity:** Maglev technology is complex and requires specialized infrastructure. The design and construction of Maglev lines requires considerable planning and technological effort.

3. Interoperability: Due to the specialized nature of Maglev infrastructure, interoperability with existing rail networks can be problematic.

Despite the disadvantages, Maglev trains are an advanced transport solution, particularly suitable for high-speed connections between distant cities. Continued research and development could help reduce costs and improve the efficiency of this technology, and perhaps one day it could be implemented to our passenger cars.

Axial Flow Motors: Optimized Performance in Lucid Air Electric Vehicles

Axial flux motors represent an innovative class of electric motors that are distinguished by their ability to generate linear thrust along the motor axis.

Operation:

Axial flux motors generate driving force parallel to the motor axis, improving torque density. Efficiency is optimized by adjusting the flow of the magnetic field.

Advantages:

1. **High Torque Density:** Axial flux motors offer higher torque density, contributing to more powerful acceleration and remarkable dynamic performance in electric vehicles like the Lucid Air.

2. **Improved Efficiency:** Due to their design, these motors can optimize the efficiency of the magnetic flux, reducing energy losses during the electrical-mechanical conversion process.

3. **Dynamic Performance:** Axial flow allows for a rapid dynamic response to changes in speed, contributing to a more responsive ride that is suitable for a variety of road conditions.

4. **Compactness and Lightness:** The axial flow configuration can be designed more compact and lighter, improving the power-to-weight ratio of the vehicle.

Detriments:

1. **Construction Complexity:** The design of axial flow motors can be more complex than previously seen solutions, resulting in additional challenges in manufacturing and maintenance.

2. **Energy Savings at Constant Speeds:** At constant speeds, axial flux motors may not be as efficient as in acceleration and deceleration situations, limiting maximum energy savings under certain circumstances.
3. **Initial Development Costs:** The implementation of advanced technologies can result in higher initial costs in both the development and production phases, affecting the overall cost of the vehicle.

Although axial flux motors offer remarkable performance, their adoption is influenced by the construction complexity and associated costs. However, with continued technological development, these disadvantages can be addressed and overcome, making axial flux motors an increasingly attractive choice for high-performance electric vehicles.

High Capacity and Power Batteries

In the vast landscape of electric mobility, batteries play a central role, representing the energy heart for electric vehicles (EVs) and shaping the future of sustainable mobility. In this in-depth exploration, we'll dive into the different battery technologies used today, focusing on crucial aspects such as weight, performance, charge density, and safety. We'll look at established technologies, such as lithium-ion, lithium iron phosphate and lithium titanate batteries, and go further to take a look at the batteries of the future.

Lithium-Ion (Li-Ion) Batteries

Lithium-ion batteries are the cornerstone of modern e-mobility. Their popularity is mainly due to the remarkable energy density they offer, ensuring a significant amount of energy stored in a relatively compact format.

Weight: One of the key advantages of Li-Ion batteries is their relatively low weight, making them ideal for use in electric vehicles, where weight management is crucial for the efficiency and operation of the vehicle.

Performance: The performance of Li-Ion batteries is remarkable, providing a wide range of voltages, energy density, and charge-discharge rates. Li-Ion technology allows you to achieve a balance between high performance and overall battery life.

Charge Density: Charge density is a key element in EV batteries, determining how quickly they can be charged. Li-Ion batteries exhibit a good charge density, allowing for relatively fast charging times compared to other technologies.

Safety: Despite their advantageous features, Li-Ion batteries can present safety risks, especially in extreme conditions or in the event of physical damage. Overheating can lead to fires or other safety concerns, emphasizing the importance of thermal management and built-in safety systems.

Lithium Iron Phosphate (LiFePO4) Batteries

Lithium iron phosphate batteries, or LiFePO4, are a variant of lithium-ion batteries, characterized by a cathode based on lithium, iron, and phosphorus. This composition gives them special properties compared to their Li-Ion counterparts.

Weight: LiFePO4 batteries tend to be slightly heavier than traditional Li-Ion counterparts, but this slight penalty is often balanced by the benefits in terms of safety and longer service life.

Performance: The performance of LiFePO4 batteries is good, providing good battery life and higher resistance to frequent charge and discharge cycles.

Charge Density: The charge density of LiFePO4 batteries is generally lower than Li-Ion, which can affect charging times.

Safety: LiFePO4 batteries are considered safer than Li-Ion. Their chemical structure reduces the risk of overheating and other safety concerns, as they do not catch fire, making them a more reliable choice in some applications.

Lithium Titanate (LTO) batteries

These batteries, also called LTO, represent another variant of lithium-ion/LiFePO4 batteries, with an anode consisting of lithium titanate.

Weight: LTO batteries tend to be heavier than the two previously seen

Performance: The performance of LTO batteries is remarkable, with an exceptionally long service life and the ability to maintain high performance in high-temperature environments.

Charge Density: The charge density of LTO batteries may be slightly lower than the other two, but a standout feature is the ability to charge extremely fast without suffering significant damage.

Safety: LTO batteries are considered safe, with increased resistance to overheating and short circuits. Their reliability and safety make them ideal for applications where safety is a priority.

Batteries of the Future

As lithium-ion batteries continue to maintain market dominance, new perspectives are opening to redefine performance parameters. Solid-state batteries represent one of the most promising directions, eliminating the traditional liquid electrolyte to offer significant improvements in safety and energy density. Research and development in this area is progressing rapidly. At the same time, graphene batteries are gaining more and more attention due to graphene's outstanding conduction properties. This technology promises higher energy density and faster charging times. While lithium-air batteries, on the other hand, exploit atmospheric oxygen as an active component in the cathode, promising an extraordinarily high energy density. However, these batteries still face significant technical challenges to become a practical reality.

The main efforts focus on the search for sustainable raw materials, advanced recycling practices and more eco-friendly technologies. The memory effect is a common problem in batteries, but modern cell management solutions are mitigating this phenomenon, ensuring a longer service life and optimal performance over time. Battery life is crucial for the cost-effectiveness and sustainability of EVs. While modern technologies are designed to handle a high number of charge-discharge cycles, actual longevity is affected by operating conditions and maintenance practices.

Chapter 6: Transmission Systems

At the beating heart of automotive engineering, where technology and a passion for driving come together, we find drive systems. These sophisticated and complex gears are at the heart of our driving experience, shaping the way we interact with the vehicle and influencing the dynamics of each journey. In this chapter, we will dive into a fascinating journey through the various powertrain technologies that animate modern vehicles, exploring their peculiarities, advantages and disadvantages.

Manual Transmission

The manual transmission represents an exciting combination of the art of driving and automotive engineering. The interaction between the driver and the gear lever creates a unique connection, especially in sports and high-performance vehicles. The presence of this transmission in these categories of vehicles is not accidental, but aims to satisfy motorists looking for an engaging and responsive drive.

An in-depth analysis of how drivers interact with the manual transmission reveals total involvement in the driving process. Every gear change becomes an act of precision and awareness, giving the driver direct control over the vehicle's performance. This mechanical dialogue creates a synergy between the driver and the vehicle, allowing for a unique personalisation of the driving experience.

The benefits of opting for a manual transmission go beyond simple tactile interaction. The feeling of being able to modulate the power of the engine through the gearbox gives the driver a sense of power and control. However, this form of transmission is not without its disadvantages. Its supposed complexity compared to automatic transmissions may require a period of adaptation for less experienced drivers. In congested urban driving situations, the constant gear change can be less practical than with an automatic transmission.

The operation of the manual transmission is based on the presence of a clutch pedal and the gear lever, allowing the driver to select the gears at his discretion. This process requires coordination between the clutch and the accelerator, offering completely different driving dynamics than automatic transmissions.

The wear and tear of a manual transmission is often related to the frequency of use and the driving mode. The clutch, in particular, as mentioned above, is subject to normal deterioration over time. However, many enthusiasts consider manual transmission maintenance to be an integral part of the experience of owning such a vehicle.

The technologies used in the manual transmission have improved over the years. From synchronizing gears to facilitate fast shifts to designing stronger clutches, modern engineering has helped optimize the performance of this transmission system.

In summary, the manual transmission represents more than just a system of gear changes. It is a driving experience imbued with emotion, highlighting the combination of technology and the art of driving a vehicle. Its presence in sports and high-performance vehicles underlines the importance of this form of drivetrain in creating an authentic bond between driver and vehicle.

Automatic Transmission

The automatic transmission, the backbone of the modern era, represents the technological evolution aimed at simplifying and making the driving experience more accessible. This innovation is the result of a long quest for practicality and comfort, revolutionizing the way drivers interact with their vehicles. With an increasingly predominant presence in vehicles in every category, from sedans to SUVs, let's now explore its influence on everyday driving.

The automatic transmission begins with its distinctive feature - the absence of the clutch pedal and the manual gear lever. This simple difference results in a smooth driving experience without active driver dynamism. The automatic transmission dynamically adapts to the driving conditions and the demands of the driver.

Practicality emerges as one of the key advantages of automatic transmission. In urban contexts, characterized by congested traffic and frequent stops, the automation of gear changes translates into less tiring driving; This is because the driver can concentrate more on his surroundings. Energy efficiency is another positive aspect, the latest technologies constantly optimize the gear ratios to ensure the best balance between power and fuel consumption. This feature results in a more economical and sustainable ride, making the automatic transmission a popular choice among those looking for efficiency without compromise.

However, despite the many advantages, the automatic transmission is not immune to some critical issues. Maintenance costs can be higher than manual transmissions, as they require more technical complexity and may require specialized intervention. In addition, some driving enthusiasts believe that the automatic transmission can take away some of the control and tactile feel associated with manual driving.

The operation of the automatic transmission is based on a complex system of planetary gears, torque converters and advanced electronics. These components work synergistically to provide the best gear changes to dynamically adapt to different driving parameters.

In terms of wear, the automatic transmission may require periodic attention, such as changing the gearbox oil and maintaining the torque converter. However, the overall lifespan can vary greatly based on the vehicle's make and model, as well as the driver's driving habits.

In summary, we can say that the automatic transmission represents a significant chapter in the history of motoring, introducing practicality and adaptability into everyday driving.

Its ubiquitous presence reflects a desire to simplify road travel, offering a convenient and efficient alternative to manual driving.

Dual Clutch Transmission (DCT)

The Dual Clutch transmission, known as DCT (Dual Clutch Transmission), represents a step forward in automatic transmission technology, introducing a faster and smoother gear change mode. This innovative solution initially became very popular in high-end vehicles, promising a smooth and continuous driving experience. Let's now delve into DCT technology, exploring its pros and cons, how it works, and the reasons for its growing success.

Agile Operation and Precise Synchronization:

At the heart of the DCT transmission lies in its ability to anticipate the driver's gear changing needs, ensuring an almost instantaneous transition between gears. Unlike traditional automatic transmissions, the DCT uses two separate clutches: one for odd-numbered gears and one for even-numbered gears. While one clutch is engaged, the other is already ready for the next change. This method eliminates downtime during gear changes.

Advantages:

The DCT offers several advantages, including extremely fast shift times, which result in more immediate acceleration and faster responses during sporty driving. In addition, the absence of interruptions in drive torque during gear changes contributes to greater fuel efficiency.

Disadvantages and Criticalities:

Significant disadvantages of dual-clutch transmission (DCT) include technical complexity, which can increase maintenance and repair costs. The initial costs of vehicles equipped with DCTs are often higher. Some drivers may perceive a lack of "mechanical feel" compared to traditional manual transmissions and notice delays in response at low speeds. In addition, city driving may be less smooth, and in some situations, such as sporty driving, the gearbox may overheat.

Success and Growing Adoption:

Despite the challenges, DCT transmission is gaining more and more popularity due to its balance of efficiency and performance. Its adoption continues to grow, prompting many automakers to implement this advanced technology in their flagship models.

Continuously variable transmission (CVT)

The Continuously Variable Transmission (CVT) is a major advancement in the field of automatic transmissions, introducing a stepless shifting approach. This technology stands out for its ability to offer an infinite range of gear ratios, aiming to maximize fuel efficiency. Let's take a closer look at the CVT, exploring its advantages, disadvantages, how it works, implementation examples, and an in-depth comparison with the Dual Clutch Transmission (DCT).

Continuous and Uninterrupted Operation:

The CVT operates through a belt or chain connected to two pulleys, with one of the pulleys connected to the input shaft and the other to the output shaft. Unlike traditional transmissions, which use a fixed number of gears, the CVT constantly varies the gear ratio continuously. This is done without interruption, allowing the engine to operate at optimal speeds in any driving situation.

Advantages:

Key benefits include increased fuel efficiency due to dynamic adaptability, a smooth ride without jerks during gear changes, automatic adaptability to driving conditions, quick response to engine speed changes, improved performance, simpler design that can reduce weight and production costs, efficiency even in hill-climb driving situations-

Disadvantages and Criticalities:

A different driving sensation that requires a period of adaptation, possible noise during acceleration, higher maintenance and repair costs, power limitations in high-performance applications, risk of overheating in extreme situations, regular oil consumption, higher initial costs and construction complexity with possible failures. These factors must be carefully considered in relation to the benefits.

In-Depth Comparison with DCT:

A comparison between the CVT and the Dual Clutch Vehicle (DCT) reveals key differences in the operating philosophy. While the DCT offers quick and precise shifts through two separate clutches, the CVT operates without defined pitches, ensuring continuous gear variation. The CVT excels in optimizing fuel efficiency and smooth driving, ideal for urban situations and quiet driving. On the other hand, the DCT is aimed at those looking for a sporty and responsive driving experience, with more immediate gear changes and greater emotional involvement of the driver.

All-Wheel Drive (AWD) and Four Wheel Drive (4WD)

The All-Wheel Drive (AWD) and the Four-Wheel All-Wheel Drive (4WD) are the spearheads in the evolution of transmissions, offering all-wheel drive to tackle different situations and road conditions. We delve into the complexity and facets of these technologies, exploring in more detail their advantages, disadvantages, maintenance considerations, wear and tear and the distinctive operation that makes them critical in various driving situations.

Operation and Differences between AWD and 4WD:

AWD, or All-Wheel Drive, operates continuously, automatically adjusting the amount of power sent to each wheel based on driving conditions. It is designed to dynamically adapt to changes in terrain or road conditions. On the other hand, the 4WD system can be manually activated or deactivated by the driver and is often equipped with a differential lock mechanism that allows power to be distributed evenly between all wheels.

Advantages of the All-Wheel Drive:

Increased traction and stability on a variety of road surfaces is the main benefit of the All-Wheel Drive. This feature results in increased driving safety, especially in adverse weather conditions or on rough terrain. The dynamic distribution of power to the wheels that need it maximizes grip, offering a higher level of control and safety to the driver himself.

Disadvantages and Mechanical Complexity:

The additional mechanical complexity can affect fuel yield, making this configuration generally less efficient than a two-wheeled vehicle. Additionally, maintenance may require a more careful approach, with additional components such as differentials needing regular verification and lubrication.

Wear and Proactive Maintenance:

As far as wear and tear is concerned, proactive maintenance is essential to ensure the proper functioning of the All-Wheel Drive. Additional elements such as axles and differentials require constant attention to avoid wear problems and ensure a reliable service life. Good maintenance can significantly extend the operational life of these components.

Semi-automatic transmission

The Semi-Automatic Transmission and Sequential Transmission represent a meeting point between the convenience of automatic transmissions and the immersive experience of manual transmissions. These systems allow the driver to intervene manually in the selection of gears, adding a touch of personalization to automatic driving, Let's examine the details of these technologies once again.

Operation:

The Semi-Automatic Transmission and therefore the Sequential Transmission share the idea of allowing the driver to manually manage the gears in an automatic context. However, substantial differences emerge in the way this interaction takes place. The Semi-Automatic Transmission uses an automatic clutch system, while the Sequential Transmission provides manual shifting of the gears without a clutch.

Advantages:

The main advantage is the flexibility offered to the driver; In fact, this system allows you to intervene manually when you want more direct control over gear selection, helping to create a more engaging driving experience. The Sequential Gearbox can offer quick shifts and precise gear selection, of course if these are allowed in the engine's operating range under certain speed conditions, all while giving the driver an additional feeling of control.

Disadvantages and Mechanical Complexity:

However, the flexibility of these systems brings with it some disadvantages. The additional mechanical complexity, particularly in the Semi-Automatic Transmission, can result in higher upfront costs than with a fully automatic or manual transmission.

Wear and Maintenance:

In terms of wear, the Semi-Automatic Transmission may require special attention to the transmission components. In the Sequential Gearbox, the care of the same could focus more on the clutch and the gear selection components. Regular maintenance is essential to ensure optimal performance and extend the operational life of these systems.

Hybrid Transmission

The Hybrid Transmission represents a sophisticated synergy between electric motors and internal combustion engines, with an operation designed to maximize the efficiency and dynamics of driving. We take a deep look at how the Hybrid Transmission works, without focusing on the traditional fuel or emissions benefits, but rather on its technical structure and the driving experience it offers.

Dynamic Operation:

The heart of the Hybrid Transmission lies in its ability to dynamically adapt to driving needs. During light acceleration phases or at low speeds, the vehicle can operate exclusively with the electric motor. In this context, the necessary power is provided by the battery or, in some cases, by a combination of the electric motor and the internal combustion engine. When more power is required, such as during intense acceleration, the internal combustion engine kicks in.

Driving Mode and Stepless Shifting:

The Hybrid Transmission features several driving modes, often included as options for the driver. This can include all-electric modes, hybrid modes that harness both energy sources simultaneously, and battery charging modes while driving with the internal combustion engine. In addition, the use of a continuous transmission (CVT) or similar systems allows for a smooth transition between driving phases, helping to maintain efficiency in a wide range of conditions

Electric Transmission

The Electric Drivetrain, a predominant feature in all-electric vehicles, embodies the transformation of the automotive landscape towards more sustainable and efficient mobility. This system represents a break with traditional transmissions, focusing on smooth acceleration and intelligent power flow management.

Power Flow Operation and Management:

Direct Electric Drivetrain is intrinsic to all-electric vehicles, where an electric motor converts electrical energy from the battery into mechanical motion. Unlike traditional transmissions, there are no fixed gears, and motion control is managed by adjusting the power supplied to the electric motor. This allows for smooth and immediate acceleration, as power can be delivered instantly without the need to go through several transmission stages.

Advantages of Electric Drivetrain:

One of the main advantages of the Electric Drivetrain is its efficiency. The absence of complex mechanical components, such as a gear transmission, eliminates the power losses associated with traditional gear changes. This results in increased efficiency of the electrical system, helping to maximize the vehicle's range. In addition, the continuous management of the power flow allows precise control of speed and acceleration, improving the driving experience.

Disadvantages and Practical Considerations:

However, the Electric Transmission is not exempt from some critical issues. The lack of a traditional transmission system can limit control options for some drivers accustomed to the feeling of shifting gears. In addition, from the point of view of wear, the Electric Transmission, being devoid of complex moving parts typical of conventional transmissions, tends to require less maintenance. The maintenance of electrical components and electronics may require the assistance of specialized technicians and specific instrumentation, and in the case of repair or replacement the costs are high.

Chapter 7: Automotive Safety

Automotive safety is a crucial priority, and technological advancement is playing a critical role in making roads safer for all users. In this chapter, we will explore in detail the different areas of innovation that are helping to improve automotive safety.

Advanced Driver Assistance Technologies (ADAS)

ADAS technologies are revolutionizing the driving experience by introducing advanced assistance and control features. Adaptive cruise control automatically adjusts the vehicle's speed to maintain a safe distance from the vehicle in front. Blind spot monitoring alerts the driver to the presence of vehicles in areas that are not directly visible. Lane Keeping Assist keeps the vehicle in its lane, reducing the risk of accidental skidding. Let's explore in general the main ADAS components present on many modern cars and their functionalities and then this AEB will be treated more specifically, this choice is dictated by the fact that they have many similarities in terms of operation, advantages and disadvantages so only one will be treated specifically for this reason.

> Adaptive Cruise Control (ACC):

- Operation: **Adaptive Cruise Control uses radar or ultrasonic sensors to monitor the distance to the vehicle in front. Based on this distance and the speed set by the driver, the system can automatically adjust the vehicle's speed, maintaining a predefined safety distance**

- Advantages: **Reduced risk of collisions, driving comfort in heavy traffic situations**

- Disadvantages: **Dependence on sensors, the additional cost and the need for proper maintenance**

> Lane Keeping Assistance (LKA):

- Operation: **The LKA uses cameras to detect lane lines. If the vehicle starts to deviate from its lane without** activating the indicators, the system can autonomously correct the trajectory or warn the driver

- Benefits: Improves vehicle stability and reduces the risk of accidents related to unintentional lane departure
- Disadvantages: Sensitivity to road conditions and the need for active driver interaction to ensure effective use

➢ Blind Spot Monitoring (BSM):

- Operation: Uses radar or ultrasonic sensors to monitor areas that are not visible directly in front of or behind the vehicle. Warns the driver of vehicles, people and objects in areas usually called "blind spots"
- Benefits: Reduced risk of collisions when changing lanes
- Disadvantages: Possible interference with other vehicles and dependence on weather conditions

➢ Tire Check (TPMS):

- Operation: System designed to constantly monitor vehicle tire pressure
- Benefits: Contributes to road safety by alerting the driver early on to pressure losses, also improving fuel efficiency and extending tire life

- Disadvantages: Need for careful maintenance and potential reporting errors

➢ Traffic Sign Recognition (TSR):

- Operation: The cameras recognize traffic signs, such as speed limits and stop signs, and show them to the driver on the dashboard or in the vehicle's display

- Advantages: Driver support in compliance with road rules

- Disadvantages: Identification errors, in adverse weather conditions or partially obscured signals

Automatic Emergency Braking Systems (AEB)

Automatic Emergency Braking Systems (AEB) represent one of the key technologies in the field of Advanced Driver Assistance Technologies (ADAS), focused on automotive safety. These systems are designed to detect emergency situations and, in the absence of an immediate driver response, automatically apply the brakes to avoid or reduce the impact of a collision. Let's see in detail how AEB works and its advantages:

Operation:

1. **Advanced sensors:** AEB systems use a combination of advanced sensors, such as radar, lidar (laser-based technology), or cameras, to monitor the vehicle's surroundings in real-time

2. **Imminent collision detection:** These sensors identify the presence of vehicles, pedestrians, or obstacles in the vehicle's path that could lead to an imminent collision

3. **Distance and speed calculation:** The system calculates the distance between the vehicle and the obstacle and evaluates relative speeds to determine the risk of collision

4. **Brake activation:** If the system believes that a collision is imminent and the driver does not take any action, it automatically activates the vehicle's brakes

Mode of intervention:

1. **Partial braking:** In some situations, the system can activate partial braking to reduce the vehicle's speed and mitigate the impact of the collision

2. Full braking: In more critical situations, AEB can activate firm braking, trying to avoid the collision completely or at least drastically reduce its impact.

Advantages:

1. Rear collision avoidance: Chain accidents can be significantly reduced, as the AEB can intervene quickly to avoid or mitigate the impact from behind

2. Pedestrian protection: Many AEB systems are designed to detect the presence of pedestrians on the road and activate the brakes to avoid pedestrian collisions

3. Impact reduction: Even in situations where a collision is unavoidable, AEB can help reduce the severity of the impact, limiting property damage and injuries

Detriments:

1. False Positives: Possible misinterpretations, causing unnecessary and potentially dangerous braking in normal situations. partially mitigated if there is an active control of the correct functioning of the sensors)

2. Dependence on Environmental Conditions: **Reduced effectiveness in adverse weather conditions, compromising the precision of the automatic braking system**

3. Complexity and Maintenance Costs: **Increased mechanical and electronic complexity, translating into higher maintenance costs and complex repairs**

Application Situations:

1. Congested traffic: **In the event of slow-moving or congested traffic, AEB can prevent or reduce impacts due to sudden braking**

2. Intersection with pedestrian crossings: **Can protect pedestrians and cyclists through detection and intervention at intersections**

3. Night driving: **Many AEB systems are designed to work even in low-light conditions, improving safety when driving at night**

AEB represents a significant step towards a safer driving environment, reducing the risk of collisions and mitigating damage in the event of an accident. Its implementation continues to grow, with more and more vehicles adopting this technology as an integral part of advanced safety systems.

As highlighted at the beginning of the chapter, many of these technologies are like each other, in fact, they all involve additional costs, maintenance in the sensors, the possibility of having false positives, but on the other hand each of these offers very significant advantages in terms of safety.

Smart Roads and V2X Communication

Smart Roads and V2X (Vehicle-to-All) Communication represent an advanced vision of road safety, where vehicles and road infrastructure are interconnected to improve safety, traffic efficiency and the driving experience. Let's take a closer look at how this works and provide some practical examples of how these technologies are revolutionizing the concept of smart roads.

Operation:

1. Street Sensors & Cameras:
    - Installation: **Smart roads have built-in sensors and cameras along the roadway or at traffic lights**

- **Detection:** Sensors detect various aspects, such as traffic flow, the presence of vehicles, weather conditions, and nearby pedestrians

2. Communication Infrastructure:

    - **Connectivity:** Smart roads are connected to a high-speed communication network
    - **Data Exchange:** The infrastructure communicates with vehicles and other connected devices, exchanging data in real time

3. V2X Communication:

    - **Vehicle-to-Everything:** V2X technology enables "Vehicle-to-Everything" communication, including vehicles (V2V), road infrastructure (V2I), pedestrians (V2P), and the network (V2N)
    - **Information Exchange:** Vehicles can exchange information with each other and with road infrastructure to improve awareness of their surroundings

4. Real-time reporting:

    - Traffic Information: **Smart roads can transmit real-time traffic information, allowing vehicles to adapt routes to avoid congestion**
    - Hazard Warnings: **Vehicles can signal emergency situations, such as accidents/obstacles on the road, to other vehicles in the vicinity**

Practical examples:

1. Smart traffic light signals:

    - Operation: **Smart traffic lights receive information about traffic density and can adapt traffic light times according to the current situation**
    - Example: **If an intersection has little traffic on one side and a lot of traffic on the other, the traffic light can automatically adjust the timing to favor the heaviest flow**

2. Accident Prevention:

    - Operation: **Vehicles and infrastructure can exchange information to prevent collisions. For example, a vehicle** might receive a notification if a traffic light is about to change while approaching an intersection

- **Example:** A car receives a warning from the traffic light that there is a vehicle crossing the intersection, allowing it to brake early

3. Real-time weather information:

    - Operation: Sensors on smart roads detect local weather conditions.
    - Example: Cars receive warnings about ice or snow on the road, allowing drivers to drive more cautiously or change routes.

4. Traffic Optimization:

    - Operation: Real-time traffic data is used to optimize vehicle flows.
    - Example: A traffic management system automatically adapts traffic lights based on the current flow, reducing wait times and improving traffic flow.

Smart Roads and V2X Communication are fundamentally changing the way we think about road safety, leading to a safer, more efficient and connected driving environment.

Impact of Autonomous Driving on Safety

Autonomous driving is a game-changer in the automotive industry. Autonomous vehicles use advanced sensors, radar, lidar, and cameras to sense their surroundings. These systems can significantly reduce accidents caused by human error, but they also present ethical and technological challenges that require special attention. Let's examine the impact of autonomous driving on road safety, considering the positives and potential challenges:

Positives:

1. Reduction of Accidents Caused by Human Error: **Most road accidents are caused by human error, such as fatigue or careless behavior. Autonomous vehicles are programmed to comply with traffic rules and react to situations in a timely and precise manner, thus reducing the risk of accidents due to human error**

2. Improved Awareness and Responsiveness: **Advanced sensors such as camera, lidar, and radar allow autonomous vehicles to constantly monitor their surroundings, detecting obstacles, vehicles, pedestrians, and road conditions. Their ability to**

process this information quickly allows for an immediate response, improving responsiveness compared to humans

3. **Compliance with Road Rules:** Autonomous vehicles are programmed to strictly comply with road rules, avoiding behaviors such as speeding, dangerous overtaking, or traffic violations. This contributes to safer and more predictable driving

4. **Adaptation to Driving Conditions:** Autonomous driving systems can adapt to driving conditions, such as rain, snow, and darkness, thanks to the combination of advanced sensors and artificial intelligence algorithms. This can improve safety in adverse conditions

Potential Challenges:

1. **Integration with Traditional Vehicles:** The gradual introduction of autonomous vehicles means that they need to be properly integrated with traditional vehicles driven by humans

2. **Software Complexity and Reliability:** Autonomous driving requires sophisticated software algorithms and neural networks to correctly interpret the surrounding environment.

The complexity of these systems increases the need for reliability and security

3. Cybersecurity Protection: With the increasing connectivity of autonomous vehicles, it is essential to ensure cybersecurity to protect vehicles from external attacks and cyber threats that could compromise road safety

4. Interaction with Humans: Communication between autonomous vehicles and human drivers, pedestrians and other road users is a major challenge. Autonomous vehicles must be able to communicate clearly and predictably to ensure safe road interactions

5. Regulations and Ethical Aspects: The definition of regulations and ethical aspects related to autonomous driving is still evolving. Issues such as legal liability in the event of accidents and ethical decisions in critical situations require in-depth attention

Advanced Driver Monitoring Systems

Recent driver surveillance systems represent a technology aimed at assessing and optimising motorists' behaviour to improve road safety. These devices use a wide range of sensors and cutting-edge technology, closely monitoring the driver's reactions. We delve into how they work, explore key components, such as cameras and radar sensors, and highlight the key benefits of these systems in increasing safety and preventing road accidents.

Operation:

1. **Driver Monitoring Camera:** A camera placed inside the passenger compartment constantly monitors the driver, detecting head movements, eyes, and other facial expressions to assess the driver's attention and alertness

2. **Pose Monitoring Sensors:** They use sensors to detect certain parameters of the steering wheel itself, the pressure on the pedals and other physical actions of the driver. They can identify signs of fatigue or lack of involvement in driving

3. **Fatigue Detection Technologies:** Specific algorithms analyze the data collected to identify signs of driver fatigue or drowsiness. If detected, they can trigger visual, sound or

vibration warnings to attract the driver's attention, in some cases they take control of the vehicle to stop its travel

4. Gaze and Eye Recognition: **Cameras can be used to recognize the driver's gaze, identifying whether they are looking at the road or other directions. This function is crucial for assessing the driver's awareness and focus**

Key components:

1. Advanced Cameras: **High-resolution cameras capture important details inside the cabin and vehicle perimeter**

2. Motion Sensors: **Detect driver movements, such as head rotation, to assess attention and awareness.**

3. Artificial Intelligence (AI) Systems: **AI algorithms analyze the data collected to identify behavioral patterns and signs of potential danger.**

4. Alerts and Alarms: **Audio-visual systems and haptic alerts can be activated in real-time to communicate critical information to the driver in a timely manner.**

Benefits:

1. Increased Road Safety: **Identifies risky behaviors or signs of fatigue, helping to prevent road accidents.**

2. Reduced Risk of Accidents: Early warnings about lane change and safety distance reduce the risk of collisions.

3. Improved Driver Behavior: Aware that they are being monitored, the driver can be incentivized to maintain safer and more responsible driving behaviors

4. Prevention of Driving Drowsiness: Detects signs of fatigue and drowsiness, triggering alerts to avoid accidents due to lack of attention

5. Safety Regulatory Compliance: Helps comply with automotive safety regulations

Advanced driver monitoring systems represent a significant step towards a safer and more informed driving experience, integrating sophisticated technologies to prevent dangerous situations and improve driving habits.

This chapter provides a detailed picture of the emerging technologies and developments that are fundamentally transforming the automotive safety landscape. The continuous adoption and evolution of these technologies will help create an increasingly safe and reliable road environment for all users.

Chapter 8: A Planet in Balance, Eco-Sustainable Vehicles and the Fight against Pollution

Welcome to the latest chapter of this journey, we will dive into the deep waters of sustainability, exploring how eco-friendly vehicles are becoming the new stars of the automotive firmament. In addition to discovering the cutting-edge technologies shaping the future, we'll delve into the crucial fight against vehicular pollution. In a world facing increasingly pressing environmental challenges, the automotive industry plays a key role in shaping the path to a balanced planet. From the explosive growth of hybrid vehicles to the paradigm shift represented by fully electric vehicles, we will explore the innovations that are transforming the way we travel.

Get ready for an adventure that combines the power of technology with environmental ethics, where every mile you ride is a statement of commitment to a cleaner, more sustainable future. Ready to start one last journey through the commitment to combat pollution from vehicles and to explore how every mobility decision can affect the course of our planet.

Environmental Impact of Internal Combustion Vehicles

Internal combustion vehicles have played a pivotal role in the history of mobility, but their impact on the environment can no longer be underestimated. These vehicles release pollutants such as nitrogen dioxide (NO_x), sulphur oxides (SO_x), fine particles and carbon monoxide (CO). According to environmental data, the transport sector is a significant contributor to global greenhouse gas emissions, with internal combustion vehicles responsible for a substantial share being increasingly targeted. CO_2 emissions, particularly linked to the use of fossil fuels, contribute to climate change and the rise in global average temperatures. In addition, cities often suffer from high levels of air pollution, with direct impacts on human health. The evolution towards cleaner propulsion technologies has become almost mandatory.

Solutions to Reduce the Environmental Impact of Internal Combustion Vehicles

In the context of growing environmental awareness and the need to promote sustainable practices, the automotive industry is faced with the challenge of making petrol, diesel and gas-powered vehicles more environmentally friendly. The transition to fully electric or hydrogen-powered vehicles is underway, but in the meantime, it is essential to explore and implement solutions that mitigate the environmental impact of traditional internal combustion vehicles. This approach reflects an awareness of the need to balance mobility with energy efficiency and environmental concerns without compromising accessibility and convenience for millions of drivers around the world. We are therefore looking at a number of innovative and practical solutions aimed at making the use of petrol, diesel and gas engines more sustainable, while promoting a future where mobility is both efficient and environmentally friendly.

Development of Sustainable Fuels

The concept of sustainable fuels represents a fundamental chapter in the ecological transition, especially in the context of vehicles that use fossil fuels to generate movement. This innovation focuses on sustainably produced fuels, reducing the environmental impact associated with the extraction, production and consumption of traditional fuels such as petrol, diesel and gas. Let's explore the stages of sustainable fuel production and the benefits of these cutting-edge techniques.

Sustainable Fuel Analysis:

1. **Renewable Feedstocks:** Sustainable fuels are based on renewable feedstocks, such as biomass, algae, or organic waste. The choice of renewable elements reduces dependence on non-renewable resources and helps mitigate the environmental impact of extraction

2. **Transformation through Bioconversion:** Bioconversion processes, such as anaerobic fermentation or pyrolysis, transform feedstocks into biofuels such as ethanol or biodiesel. These processes emit significantly fewer greenhouse gases than the production of conventional fuels

3. Production of Synthetic Fuels: **Some approaches involve the production of synthetic fuels, such as e-diesel or e-gasoline, using electricity generated from renewable sources. This step helps reduce the overall carbon footprint of the fuel**

4. Advanced Refining: **Advanced refining processes, including catalysts and purification technologies, are employed to obtain sustainable fuels with quality specifications comparable to those of traditional fuels**

Advantages of Sustainable Fuels:

1. Reduction of Greenhouse Gas Emissions: **Sustainable fuels, derived from biomass / renewable electricity, contribute to significantly reducing greenhouse gas emissions**

2. Less Dependence on Fossil Fuels: **The use of renewable raw materials reduces dependence on fossil fuels, contributing to greater energy security and diversification of energy sources**

3. Utilization of Existing Infrastructure: **Sustainable fuel production is designed to integrate into existing infrastructure, allowing for a gradual shift to more sustainable technologies.**

4. Incentivizing the Circular Economy: **The use of organic waste or biomass helps to promote the circular economy,**

exploiting materials that could otherwise be destined for landfill.

5. Compatibility with Existing Vehicles: Sustainable fuels are designed to be compatible with existing internal combustion vehicles, reducing the need to replace entire fleets of vehicles.

6. Environmental and Social Sustainability: The production of sustainable fuels can promote sustainable agricultural practices and create economic opportunities in local communities.

Investing in the development of sustainable fuels is a crucial step towards making the automotive sector more sustainable, ensuring a smooth and realistic transition to greener energy sources. The challenge now is to implement these solutions on a global scale, one step at a time, for a future where mobility and sustainability coexist.

Emission Reduction Solutions

The implementation of advanced emission reduction technologies, such as more effective catalytic converters and direct injection systems, helps to minimize the environmental impacts of internal combustion vehicles.

Three-Way Catalysts (TWCs):

How it works: The TWC is made up of three main components: the oxidation catalyst, the reduction catalyst, and the storage catalyst. These components work synergistically to convert pollutants into "safer" substances.

Details: The oxidation catalyst converts CO and hydrocarbons into CO_2 and water. The reduction catalyst reduces NOx to nitrogen and water. The storage catalyst reduces emissions during the cold start phase.

Application: Mainly used in gasoline engines, it finds wide applications in the exhaust systems of internal combustion automobiles, light commercial vehicles, hybrids, plug-ins, and natural gas vehicles.

Diesel Particulate Filter (DPF):

Operation: The DPF captures the solid carbon particles present in the exhaust gases, preventing them from reaching the atmosphere.

Details: The filters can be porous wall or ceramic with catalytic coating. Regeneration occurs through the combustion of trapped particles.

Application: Applied in diesel engines, such as cars, vans, trucks, and commercial vehicles, DPF is essential for capturing and reducing solid particles present in diesel exhaust gases. In addition to the automotive sector, it is used in industrial and agricultural machinery, railway vehicles, marine vessels and diesel generators.

Selective Catalytic Reduction (SCR):

Operation: The SCR uses a catalyst and a urea injector (AdBlue is a solution of urea and water)

Details: Urea is injected into the exhaust gas, reacting with NOx in the catalytic converter. The catalyst reduces nitrogen oxides into nitrogen and water

Application: The Selective Catalytic Reduction (SCR) system and the Diesel Particulate Filter (DPF) are often used in combination in diesel vehicles. While the DPF retains solid particles, the SCR reduces nitrogen oxides (NOx). This synergy allows for complete management of emissions

Exhaust Gas Recirculation (EGR):

Operation: The Exhaust Gas Recirculation (EGR) system operates by reducing the combustion temperature, by reintroducing a part of the exhaust gases into the intake. This practice reduces the amount of oxygen available, thereby controlling the temperature during combustion and mitigating the formation of nitrogen oxides (NOx)

Details: There are two main categories: mechanical and electronic EGR. Mechanical EGRs use mechanically controlled valves to manage the flow of exhaust gases, while electronic EGRs employ electronic actuators and advanced sensors for more precise and dynamic management. The latter allow finer control of exhaust gas recirculation according to driving conditions and engine needs

Application: It reduces NOx emissions especially in diesel engines, but also finds some applications in other internal combustion engines, this reduction is done upstream, so as to lighten the downstream load in the SCR systems described above

Environmental Benefits of the Analyzed Technologies:

1. Compliance with Regulations: **They contribute to compliance with environmental regulations and emission restrictions.**

2. Improved Air Quality: **Reduce pollutant emissions, improving air quality and human health.**

3. Application Versatility: **Many of these technologies can be implemented on several traditional engines.**

The EGR, DPF and SCR systems described above have many disadvantages that affect performance, consumption and additional costs. EGR, by recirculating exhaust gases, can accumulate carbon deposits, compromising engine efficiency and fuel consumption. The DPF, requiring periodic regeneration, can increase fuel consumption and cause reliability problems. SCR, using additives such as urea, involves additional costs and requires a dedicated tank. In addition, these systems can lead to an increase in the complexity of the engine and its components.

Vehicle Weight Reduction

Reducing vehicle weight is a key goal in the modern automotive industry, as it directly affects fuel efficiency, performance, and environmental impact. The use of innovative materials contributes not only to reducing the overall weight of the vehicle but also to improving its eco-friendliness. Let's explore some of the materials used and their impact on the environment.

Aluminum Alloys:

- Features: Lightweight and durable, they offer excellent corrosion resistance. They are often used in structural components and bodywork
- Eco-friendliness: Mining and processing aluminum requires energy, but recycling it is efficient, helping to reduce environmental impact

High Strength Steel:

- Features: Offer superior strength to lighter weights. They can be used to reinforce crucial parts of the vehicle structure
- Eco-Friendly: Using recycled or sustainably sourced steel can reduce environmental impact

Carbon Fibers:

- **Features:** Lightweight and incredibly strong, carbon fibers are often used in high-performance components such as body parts
- **Eco-Friendliness:** Carbon fiber production requires energy-intensive processes, but technologies are evolving to reduce environmental impact

Biodegradable plastic:

- **Features:** Biodegradable plastic is used for non-structural components
- **Eco-Friendliness:** Helps reduce the problem of plastic waste, especially when derived from sustainable sources.

Wood and Natural Materials:

- **Features:** Materials such as wood can be used in interiors and non-structural components to reduce the use of traditional materials
- **Eco-Friendly:** Sustainable when sourced from responsibly managed forests

Benefits of Weight Reduction:

1. **Fuel Efficiency:** Reduces fuel consumption, increases the number of kilometres travelled and decreases expenses over time

2. **Performance:** Contributes to higher performance, improved acceleration and improved handling

3. **Reduced emissions:** Minimizes greenhouse gas emissions thanks to a lighter load

4. **Eco-Sustainability:** the practice of recycling contributes to a further reduction of environmental impact

Balancing weight reduction and eco-friendliness requires a holistic approach, considering the vehicle's production, use and end-of-life. Continuous innovations in materials and manufacturing practices are pushing the industry towards lighter and more environmentally sustainable vehicles.

Detailed Comparison in Emissions, Resource Consumption and Sustainability

Exploring and comparing electric vehicles (EVs), hybrids and hydrogen-powered vehicles is a complex journey, intertwining with ever-evolving challenges. Environmental, political and public opinion considerations play a crucial role in this scenario. To comprehensively address this complex challenge, it is essential to carefully examine each stage of the vehicle lifecycle, from production to disposal, to fully understand the impact of these technologies.

We start by exploring emissions during use, a crucial aspect for assessing the immediate impact of vehicles on air quality and climate change. EVs emerge as the undisputed leaders in this space, providing a zero-emission driving experience on site. Numerous studies and field trials indicate that, compared to internal combustion vehicles, EVs significantly reduce the environmental impact during the operational phase. This becomes particularly evident in areas where electricity is produced from renewable sources, helping to further reduce the overall ecological footprint.

Hybrid vehicles, on the other hand, represent a compromise, reducing emissions while driving thanks to the integration of an electric motor, but maintaining a continuous dependence on fossil fuels. Government policies and initiatives to promote low-emission technologies are crucial to driving the success of these solutions.

The debate on the availability of resources intensifies when examining the propulsion technologies of each vehicle. EVs, powered primarily by lithium batteries, face the challenges of extracting materials. Mining ethics and the environmental impacts of mines become issues of concern that require responsible and sustainable solutions.

On the other hand, hydrogen-powered vehicles face challenges related to the use of platinum in fuel cells. Platinum is an expensive material with limited availability, raising questions about its long-term sustainability. The search for more sustainable alternatives and the adoption of responsible production practices therefore become imperative.

The validity of the claims about the efficiency and sustainability of the technologies emerges through field tests and comparative analysis. Battery endurance studies in EVs, for example, provide critical data on durability and the ability to maintain high performance over time. Range testing and evaluations of hydrogen cells provide vital insight into the practicality of these technologies.

A tangible example of progress is the continuous evolution of lithium-ion batteries. Improvements in energy density and durability are helping to make EVs increasingly accessible and attractive to the public. Extensive testing shows the steady growth in performance and cost reduction, positioning EVs as increasingly competitive options in the market.

In conclusion, the detailed comparison between electric, hybrid and hydrogen-powered vehicles requires a critical and comprehensive exploration of the various stages of the life cycle. The convergence of data from field tests, life cycle analysis and ethical considerations is essential to gain an overall perspective on the sustainability of each technology.

In the era of sustainable mobility, the balance between impacts and benefits must be carefully assessed. Continuous innovation and the adoption of more sustainable practices are essential to shaping the future of mobility in a responsible and environmentally harmonious way. We are in a critical phase where every choice counts, and the automotive industry plays a fundamental role in defining a sustainable path towards the future of mobility.

Conclusion

On this fascinating journey through the depths of modern automotive mechanics, we immersed ourselves in the beating heart of contemporary vehicles, exploring the intricate operation, essential maintenance and revolutionary decisions in terms of propulsion technologies. Four-wheelers, a symbol of freedom and progress, continue to move forward in time, driven by an insatiable desire for innovation and sustainability.

Each chapter of this book is a stage in an exciting journey, from the birth of the first mechanical motions to the present day, in the field of automotive mechanics, challenges continue to emerge that are faced with different adaptations and overcomings of usual habits, through the complex structure of modern vehicles, from the impetuous engine under the hood to the sophisticated transmissions, We have explored every component that, working together synergistically, makes our daily journey possible.

We approached maintenance as an act of love towards our faithful companion on four wheels, ensuring that every component works in unison to ensure optimal performance and extended life. We have learned that engine care is more than a practice, it is a commitment to longevity and efficiency.

In the following chapters we catapulted into the revolution of power systems, from the classic internal combustion systems to the new frontiers of electric innovation. We explored the various powertrain, drivetrain and road safety options, unveiling the paths to sustainability we are taking and those we still need to explore.

Finally, at the end of this journey, we faced the challenge of vehicular pollution, questioning our environmental impact. With our eyes turned to the future, we looked at eco-friendly alternatives, from the growth of hybrids to the exciting electric revolution.

Welcome to a world of knowledge, power and boundless driving, where the evolution of automotive mechanics embraces the future with determination and vision. May this journey be as instructive as it is exciting. Enjoy the reading!

www.ingramcontent.com/pod-product-compliance
Lightning Source LLC
Chambersburg PA
CBHW020424220526
45464CB00002B/554